How to Write and Publish

Engineering Papers and Reports

Second Edition

The Professional Writing Series

This volume is one of a series published by ISI Press®. The series is designed to improve the communication skills of professional men and women, as well as students embarking upon professional careers.

Books published in this series:

Communication Skills for the Foreign-Born Professional
  by GREGORY A. BARNES

The Art of Abstracting
  by EDWARD T. CREMMINS

How to Write and Publish a Scientific Paper
  by ROBERT A. DAY

A Treasury for Word Lovers
  by MORTON S. FREEMAN

The Story Behind the Word
  by MORTON S. FREEMAN

Presenting Science to the Public
  by BARBARA GASTEL

How to Write and Publish Papers in the Medical Sciences
  by EDWARD J. HUTH

How to Write and Publish Engineering Papers and Reports
  by HERBERT B. MICHAELSON

How to Write Papers and Reports about Computer Technology
  by CHARLES H. SIDES

How to Write a Usable User Manual
  by EDMOND H. WEISS

How to Write and Publish

Engineering Papers and Reports

Second Edition

Herbert B. Michaelson

iSi PRESS®

Philadelphia

Published by

iSi PRESS ®A Subsidiary of the
Institute for Scientifc Information®
3501 Market Street, Philadelphia, PA 19104 U.S.A.

© 1982, 1986 ISI Press

Library of Congress Cataloging in Publication Data

Michaelson, Herbert B., 1916–
  How to write and publish engineering papers and reports.

  (The Professional Writing Series)
  Bibliography: p.
  Includes index.
  1. Technical writing.   I. Title.   II. Series
T11.M418   1986   808'.06662   85-30512
ISBN 0-89495-055-x
ISBN 0-89495-056-8 (pbk.)

Printed in the United States of America.
93 92 91 90 89 88 87 86   8 7 6 5 4 3 2 1

Contents

Preface to the First Edition

There is no simple solution to the deep-seated problems of engineering writing and publishing. My approach to helping the engineering student and the practicing engineer is to divide the problems into manageable pieces in the chapters of this book.

From my experience as an editor and author of engineering papers, I decided to devote the text to several practical aspects of technical communication.

- The first is the basis for sound engineering writing: logical organization, sensitivity to audiences, and lucid, convincing expression of ideas.
- The second is an important role of the engineer in the professional community: publication of the manuscript.
- The third and most decisive aspect is the common thread in the fabric of all writing and publishing: the engineer's strategic choices.

Accordingly, the book is organized along these lines. The first three chapters discuss the objectives and motivations for communication. Chapters 4 to 17 explain the techniques of planning and writing. Later chapters give methods of publication and presentation. A pervasive topic throughout is the author's role as strategist, which I summarize in the last chapter.

The scope covers formal writing for technical audiences—manuscripts on research, development, or analysis of engineering topics. Examples show the distinctions among papers for journals and conferences, articles for trade magazines, internal company reports, and theses and reports for the academic community.

My personal philosophies about writing and publishing are reflected in every chapter. One is my belief that an author's choice of sound tactics and strategies not only enhances the quality of the manuscript and ensures its ultimate success, but also reduces the effort required to write it. A second belief, expressed especially in Chapter 10, is that the engineer should emphasize the strong points and subordinate the peripheral information in the manuscript. I believe, too, that most writing about engineering developments is creative effort, subtle but powerful. In Chapter 5, for example, I propose a writing method that helps the engineering work, and I suggest that an author's insight is a two-way bridge between the manuscript and the work.

I have attempted to fill the needs of engineer authors by showing how to anticipate problems before they arise. One example of this is my solution to the delicate question of choosing co-authors. Another is a way of scheduling the document for completion at the most opportune time. Of importance to all authors is the role of constructive criticism in the writing and publishing process. In addition, there are techniques for forestalling the negative criticism of reviewers and, if these methods fail, of defending the refereed manuscript.

Included also are my favorite suggestions for using writing as a stimulus to both reader and writer. And I propose in one chapter that composing a manuscript on a personal computer is an aid to writing and thinking processes.

My thinking is that writing and publishing are a part of an engineer's professional development that bring their own rewards—recognition for technical achievement and the joy of effective expression. In retrospect, I might have entitled this book *The Joys of Engineering Writing.*

Herbert B. Michaelson
April 1982

Additions to the Second Edition

The first edition emphasized techniques for sound writing and successful publishing that aid the *professional development* of the engineer. In the present edition I decided to add several chapters to serve the same purpose.

Two new chapters are intended for the engineering student. The first deals with important decisions that must be made in preparing a master's thesis or a doctoral dissertation. Unlike the standard instructions for fulfilling degree requirements, my advice is to plan very early in the process for future publication in a journal. Chapter 17 shows specifically how to meet the challenges of the faculty committee and at the same time plan for peer recognition by publishing professionally. The second new chapter deals with the student's use of a personal computer for composing manuscripts. Chapter 12 suggests that quick, neat corrections are only a small part of the benefits. Interaction with the machine provides some outstanding advantages that eventually become evident in the quality of writing and in the quantity of output.

I also wrote two new chapters for the practicing engineer. One addresses the pervasive problem of confidentiality in R&D work. Many engineers cannot publish their results promptly in journals because of proprietary restrictions in industry or security regulations in defense work. In Chapter 18 I show how a prospective author can take the initiative— in strictly legitimate ways—to ensure early publication in the journals.

The second new chapter for engineers may at first glance seem a bit out of place in this book. It is called "How to Review Engineering Manuscripts." I decided to write this chapter, however, for important reasons in keeping with the main thrust of the book. An engineer who writes and publishes will gradually become known for contributions to the state of the art and can be approached by a journal editor to review a manuscript. Chapter 22 suggests that appraising new work offered by others goes far beyond nitpicking or fault finding. Offering a fresh point of view and perhaps new insights into the engineering results can be a contribution in itself and a stimulus for the reviewer's own professional development.

I am especially grateful to Professor Edward R. Ernst, Dr. Rudy J. Joenk, John A. Lacy, Dr. Leighton E. Sissom, and Q. William Simkins for invaluable help and advice in reviewing the newly added material. I also thank editors Maryanne Soper and Estella Bradley for excellent work in preparing the manuscript for publication.

Herbert B. Michaelson
November 1985

Acknowledgments

This "how to" book was written at the suggestion of Robert A. Day, author of a well-known textbook on writing scientific papers. The ideas in my book are derived from some of my published papers on the subject, from the writing seminars I conducted for several years at Yale University, and from my extensive experience as associate editor of the *IBM Journal of Research and Development.*

I have a deep sense of appreciation for the in-depth discussions with my colleagues at IBM, including Gene Damm, Jeff Hibbard, John Lacy, and George Stierhoff. In addition, I acknowledge the valuable suggestions on portions of the manuscript offered by Wally Brooks, Al Davis, Rudy Cypser, Vingie Merritt, Ralph Preiss, and Pete Vrouvas.

For helpful criticism and advice on the manuscript, I express my thanks to Dr. Rudolph J. Joenk, Jr., Robert A. Pease, and Ben H. Weil. I also thank the following, who have read and commented on portions of the book: Professor Michael J. Driscoll, Massachusetts Institute of Technology; Professor Edward R. Ernst, University of Illinois; Dr. Jack Selzer, Pennsylvania State University; Dr. Terry Skelton, University of Delaware; Dean Leighton Sissom, Tennessee Technological University; Dr. Mary Fran Buehler and Helen Caird, Jet Propulsion Laboratory; Dr. Frank Smith, McDonnell Douglas Corporation; Harold F. Osborne, consultant; and Lola Zook, consultant.

I am grateful to my department director, Bill Simkins, for encouragement and support. And I owe a special thanks to Kathy Zoll for skillfully and diligently preparing the text-processed manuscript.

Chapter 1

How to Define Quality
in Engineering Manuscripts

The Importance of Quality

An engineer's work is never quite complete until he or she has described what was accomplished, by writing a technical report for an engineering organization or submitting a paper for journal publication. In either case, the quality of the manuscript reflects on the character of the author's work and reputation.

Quality, of course, is inherent in any conceivable kind of engineering effort, and the merit of an engineer's work can be measured according to the standards of the profession. The quality and value of the manuscript, however, are less easy to define—the definition will depend somewhat on the reference frame of the reader.

If you are to write a manuscript that will be known for its quality, you must first estimate how your readers will judge it, even though viewpoints may differ widely. For example:

- *Readers of reports or journal papers* are mainly interested in new or definitive technical information, presented in an understandable fashion at an appropriate technical level.
- *Editors and publishers* consider a manuscript in terms of how it will fit the subject areas of their publication, e.g., the technical interests of the readership, and of whether it meets editorial standards of style and content.
- *The author* views it in terms of the character of personal accomplishment and how it might affect his or her professional reputation.

Even though various classes of readers have differing criteria, your manuscript can satisfy them all if it has certain basic elements of quality.

1

The Elements of Quality

You can best approach the question of quality, then, by considering the viewpoints of potential readers. A nontechnical reader, for example, may believe that clear expression is the first essential, and much has been written about the virtues of clarity and the sins of ambiguity. But for the engineering reader, literary style alone (which includes conciseness, objectivity, and good diction) does not ensure a high-quality manuscript if the technical results are not sound and the conclusions are not valid.

A realistic view of the quality of a manuscript must include all of its elements: technical content, organization, literary style, validity, and significance. These elements tend to overlap and to interact with one another, but a brief discussion of each is revealing about the nature of a quality manuscript.

Content

The best way to write an engineering manuscript is to build quality into it during the planning rather than to try to inject quality into it after the paper is finished. The heart of the matter, and the very first consideration, is its technical content. On an overall scale of quality, content will certainly rank higher in importance than either style or organization of the paper. In the initial planning you must make a value judgment: Will the technical content be sufficient to justify the time and effort necessary to write a manuscript?

In answering that question, you must be guided by the intended purpose of the manuscript. If it is to be submitted to a journal, will the subject be suitable for a particular publication and its readers? If instead it is to be an internal report, will it be written mainly as a matter of record, intended for hypothetical users who might need it to avoid duplication of work? Is your report intended to provide a useful understanding of a device, process, method, or system? Or will it be an analysis of a problem?

It is unfortunate that sometimes an engineer will prepare a manuscript without carefully considering its real purpose. For the reader, such a manuscript does not whet the appetite—it is neither fish nor fowl and has no identifiable flavor.

Therefore, before even attempting to organize your ideas for a paper, you should ask yourself the following questions:

- What class of readers would find it most useful? Why?
- Is the work original? If so, what is actually new?
- What portions of the work are the best topics to include?
- What portions would detract from the quality of a manuscript?

- How do you characterize its content? Is it conceptual, developmental, analytical, or descriptive?

You do not have to present a new engineering concept in order to write a manuscript of quality. A purely descriptive report about the features of a well-known device or system, for example, can be a superior manuscript, of wide interest and utility, and important for tutorial purposes and for its archival value. But whatever the character of a manuscript, identifying your motivations at the start will help to organize the writing project.

Organization

The best structural qualities of the writing are those that emphasize your main purpose. The usual difficulty with a poorly organized manuscript, in which the author's emphasis is out of balance, is that the reader cannot see the woods for the trees. In constructing a manuscript of high quality, you must choose and arrange several sections in a way that shows your results to best advantage and indicates why they are important.

Most engineering papers begin with a brief abstract. This is followed by an introductory section that indicates the motivation for the work, gives background information on previous efforts, and describes the nature of the engineering problem and the author's approach to solving it. For the rest of the manuscript, the key to the best structure lies in the choice of a main theme for one section and of the supporting material for lesser sections, suitably de-emphasized.

The most prominent section, for example, might be devoted to any of the following themes:

novelty of design method	improved appearance
new principle of operation	better quality control
improved operating	longer life
characteristics	higher efficiency
lower cost	new analysis or
ease of use	interpretation

This section of the manuscript should be supported by illustrations, tables, or examples that show your main points and chief contributions.

This scheme for organizing a paper is an extension of the conventional principles for constructing effective sentences and paragraphs, i.e., proper emphasis and subordination of ideas. One more hallmark of quality is a concluding section that not only summarizes the results but also their significance in relation to prior published work. A good style of organization thus brings out the best of the technical content.

Literary Style

An eternal truth that emerges from any class in English composition is that "writing is the man himself." It is also true, however, that in any report of engineering development the writing reflects the work itself. The character of an engineer's efforts, particularly if they are in the area of design or development work, inevitably finds its way into the style of the manuscript. Although these effects are subtle, they can exert a powerful influence on the writing. For example, an engineering project that is dull and routine tends to be described in a dull, plodding report. Development work full of innovation and exciting progress tends to inspire the author to write an interesting and readable manuscript.

Beautifully clear expression can be motivated, for example, by the beauty and originality of a new engineering design. Logical and orderly exposition is frequently related to the logic and order of the engineering process it describes. The converse is equally true: confused thinking in the thread of development work is difficult to weave into the fabric of logical, convincing prose.

Another important property of style is readability. The several types of readability scales, which were reviewed recently by Barry,[10] are usually based on word length and sentence length. This approach may be valid for certain types of literature. In papers written by engineers for their peers, however, readability depends more on the proper use of technical language and on the full development of sentences and paragraphs. Such development shows relationships among technical ideas. These logical connectives tend to be lacking in short sentences. For this reason, short words and sentences are not necessarily a stylistic advantage for the engineer.

Clarity and readability are by no means the ultimate in literary style. A clearly written, readable manuscript may have other defects that limit its usefulness to readers and even its suitability for publication. Conciseness, for example, is one aspect of style that becomes increasingly important as today's extensive engineering literature grows in volume. The long-winded paper or report, full of excess information, may be abundantly clear but too demanding of a reader's time and attention. The typical engineering reader is deluged with reports, papers, and magazines. Keeping up with the literature is difficult, even in the engineer's own special field of work. For the busy reader conciseness is an outstanding virtue. Nevertheless, there is a sharp distinction between conciseness and brevity, as pointed out by Strunk and White.[57] The concise paper is full of substance that pertains to its main theme; it says much in few words. A manuscript that is merely brief may or may not have the important quality of conciseness.

Other elements of style that have a potent influence on readers are diction and sentence structure. Developing mature judgment in the choice of technical language is especially important for the engineer author; each branch of engineering has its own terminology and each discipline has its own shoptalk and jargon. In the choice of technical language for formal writing, an author must consider what is appropriate for the readers.

The most powerful effect on technical style is probably that of sentence structure, which is the prime vehicle for emphasis, subordination of ideas and variety of expression. An abundance of simple declarative sentences creates a choppy style, full of vigor and directness but lacking in the qualifications and relationships that are so characteristic of engineering writing. Too many compound sentences tend to present ideas in balance and in contrast without the use of modifiers and qualifiers. A preponderance of complex and complex-compound sentences tends to overqualify and gives the reader the impression that the author is hedging on every statement. The best style, of course, is a balance of the various types of sentences. As pointed out by Weisman,[64] professional writers use about 50 percent complex sentences, 35 percent simple sentences, and 15 percent compound and other types.

Literary style, then, affects the meaning, flavor, and general effectiveness of engineering writing. The ultimate in effectiveness, however, lies not in the style of writing, but in the way the technical results are shown to be valid and significant.

Validity and Significance

If your paper offers any kind of new results, you have a special obligation to your readers to show that the information and findings are valid. Your claims are more convincing, of course, when they are amply supported in the manuscript. One way to establish credibility is to cite the alternative methods of solving your engineering problem and to give the reasons for your choice. Another is to mention the negative aspects of your results. Showing their limitations can be a strength instead of a weakness. To indicate what your design can do and cannot do is a far better strategy than to confine your discussion to claims of superiority.

A superior design is best understood and appreciated when it is described in a clear and concise style. But clarity and conciseness are no substitute for the integrity of an engineering concept or design. For example:

> The truss structure of this completed bridge, in combination with the use of Superalloy 10-5-85 carbon steel, permits a maximum vehicle traffic load of 75 tons.

Quite clear. Specific and unmistakable. But if a 60-ton traffic load should cause collapse of the bridge, the author's statement would not be valid.

Some authors assume that describing their engineering developments in a paper is sufficient and consider a discussion of the significance superfluous. If the engineering work is worthwhile, won't the beauty of the results speak for itself? The answer to that question lies in the competitive nature of modern engineering literature. Not all of the interested readers will leisurely read every word of the paper, and any busy reader will recognize the value of at least a short section discussing the implications of the work. The quality of a manuscript is thus always enhanced by a brief recapitulation of its significance. This summary might mention the critical factors that led to the success of the project, its relation to other published work, the potential applications of the results, and the directions of future developments.

Summary

All of the elements of quality contribute separately to the overall merit of an engineering manuscript. Inadequate content cannot masquerade under the cloak of fine writing. Nor can outmoded engineering concepts hide behind a facade of logical structure and orderly presentation of ideas. Literary style must be combined with effective organization, and results need to be fortified with the kind of balanced intepretation that identifies a manuscript of quality.

Chapter 2

How to Match Your Objectives
with Reader Interests

Empathy for Your Readers

The author of an effective engineering manuscript has a subtle and intriguing quality known as empathy. Indeed, the ability to put yourself in the reader's place and to understand the reader's needs and interests will make all the difference between writing an informative (but uninspired) paper and writing one that attracts and influences its audience. The principles of audience analysis are described by Pearsall[50] and Mathes and Stevenson.[37]

There is a certain mystique about an author's feeling for his readers, and it is not always easy to see where that sensitivity shows up in the manuscript. Empathy is part of an author's insight that connects the audience with the author's work and is a driving force that contributes to the quality of the writing.

This aspect of empathy can be understood by considering an example of how an author who senses the needs of the audience can subtly modify the manuscript to meet those needs. Suppose that, in a paper on the design of a refrigerating plant for cold storage, the author includes what is ordinarily expected—a brief statement of the main purpose—somewhere in the early part of the paper:

> The purpose of this paper is to show the design of a refrigerating plant in which a novel layered structure of foam Insulon, instead of the conventional glass wool, is used as insulation in the wall of the cold storage room. The use of this new insulator drastically reduces heat gain in the storage space and thus lowers the electric power consumption of the compressor motor by nine percent.

Suppose, instead, the author knows that the paper will be published during

7

a period of severe drought. Under these circumstances the following sentence could be added, which otherwise might have been omitted:

> The condenser in this refrigerator is the well-known air-cooled Model 54B, which saves the water consumption of 20 gal/min that would have been required by water-cooled types.

The author thus combines a statement of objectives with a reader-oriented statement that had not even been a part of the design problem.

You may well object that, although empathy is obviously desirable, you cannot possibly develop such insights if you don't know who your readers are. The question of identifying your readers is discussed later, but first we should consider the all-important matter of your objectives.

Your Objectives and Motivations

The conventional way to begin any engineering paper or report is first to state the engineering problem and then to show what has been done to solve it. This statement of the problem is generally the same as the objectives of your engineering project. However, those objectives, *the purpose of the work,* are frequently quite different from your motivations for writing, *the purpose of the paper.* To reach your readers effectively, you must have a clear understanding of the difference between those two purposes. Only then can you match your objectives with your readership. If you confuse the purpose of the work with the purpose of the paper, your manuscript will lack focus and direction.

We can shed a little light on the understanding of your purposes by examining some of the usual objectives of engineering work and comparing them with a list of personal motivations for writing. First, let us list some of the broad purposes of engineering effort:

YOUR OBJECTIVES
> To develop a new theory or principle
> To show practical applications of known principles
> To develop a solution for an engineering problem in a device, material, system, or process
> To design a new structural form
> To develop a new or improved method
> To establish a set of standards

Now contrast these with some of the various reasons for writing:

YOUR MOTIVATIONS
> To be known for your work and accomplishment

To be published in a prestigious journal

To attend a conference, meet with your peers, and publish your paper in the proceedings

To record your developments for their archival value

To make a progress report of ongoing developments in your engineering project

To propose a new program

To describe, for tutorial purposes, certain engineering developments or principles

To establish precedence by publishing before your competitors can

To satisfy requirements for an academic degree

The particular work objectives and personal motivations that apply to your manuscript will strongly affect its structure and flavor. For example, if your objective is to present a theory, you would have an analytic section and probably one on the experimental verification. If, on the other hand, your objective is to describe an application of a well-known principle, your paper would be constructed along different lines, probably describing the method of application, its advantages, and its usefulness.

The manuscript will also be affected in special ways by your motivations. If, for example, you plan to publish an original piece of work in a journal, much of the emphasis will be on how your design or system differs from previous types and why it is better. If, instead, your motivation is to write a technical report for use within your organization, you might be describing only engineering features and their utility. The effects that your main purpose has on the planning of your manuscript are discussed in Chapter 4.

But no matter how important your objectives and how valid your motivations, the message in your paper will not get through to your readers unless it somehow matches their interests and technical level. And your message will not always seem important to the audience unless it relates in some way to their own objectives.

Your Readers' Needs and Interests

How can you know who your readers will be, and how can you judge their actual needs and interests? There are several ways in which you can make a good estimate of your readership. Before you can do this, however, you must decide on a vehicle for your manuscript. The initial step—deciding where your manuscript is to appear—is the key to understanding your readers.

If you are writing an engineering manuscript for your peers, it will probably take one of the following forms:

- *A paper* for a journal of one of the professional societies
- *An article* for a technical trade magazine
- *An oral presentation* for an engineering conference or seminar, with subsequent publication in its Proceedings
- *An internal report* for use by your engineering organization
- *An external report* of work done by your organization
- *An academic thesis* for fulfillment of a degree requirement

These six kinds of publications have separate, fairly well-defined classes of readers. If you can pick your target while planning your paper, you can then take careful aim at your objective: their interests.

Suppose you pick your target after you finish the manuscript. You could (and many authors do) prepare your manuscript in a closet, as it were—oblivious of a future audience—and then venture with it to the outside world, hoping to find readers. Under those circumstances you may gradually discover your mistake and learn that it is far easier to design your manuscript to fill a need among readers than it is to create a need after your paper is written.

You will be much better off if you choose a vehicle first and consider its particular class of readers. If, for example, you are to report the results of original research or development that could be published in a journal, look at the publications in your field of work. Journals frequently identify their readers' interests, e.g., the kinds of papers they accept for publication. Some professional society journals, such as *Communications of the Association for Computing Machinery,* include a definitive statement of their scope and purpose on the table of contents page. Others, such as *Proceedings of the Institute of Electrical and Electronics Engineers,* invite authors to get in touch with the editor before the manuscript is submitted. The following appears on its masthead:

> The *Proceedings of the IEEE* welcomes for consideration 1) contributed tutorial-review papers in all areas of electrical engineering, and 2) contributed research papers on subjects of broad interest to IEEE members. The prospective author of a tutorial-review paper is encouraged to submit an advance proposal giving an outline of the proposed coverage and a brief explanation of why the subject is of current importance and of his relation to the subject.

Most professional societies also provide pamphlets of instructions to authors.

Another clear indication of the scope of reader interest is the table of contents of the journal. Look through several issues to get a feeling for the subject matter and the technical level of the papers. When you have selected a journal with a coverage that seems closest to your own topic, you have found a target. One way to find the bull's-eye is to ask the editor of the journal if the subject and range of content of your forthcoming manuscript seem appropriate. Prior agreement with an editor is no guarantee that your paper will be accepted, but it goes a long way toward defining the interests of readers.

In addition to journal papers, five other kinds of articles and reports were listed above. For these you can find similar definitions of readership.

The trade magazines, for example, have a strong concern for their audience. *Quality* (a magazine of product assurance) suggests in its "Guidelines for Authors" that there are six categories of content for maximum reader interest in every issue. These are

Test and data analysis systems
Quality and reliability engineering
Manufacturing and process quality control
Management for performance improvement
Control of purchased material
Inspection and evaluation

For each of these categories, the guidelines offer a helpful subject breakdown.

Similarly, when a call for papers is issued for an engineering conference, it usually cites the range of subjects and the desired emphasis. The *International Symposium on Automotive Technology and Automation*, for example, gave this advice to authors:

> The subjects for papers will cover the whole range of automotive technology and automation, including automation of manufacturing processes, component testing, testing for development, production testing and diagnostics, quality control, electronic engine control, computer aided engineering and manufacturing, sensors, emissions, lubricant testing, fuel economy and safety. PARTICULAR EMPHASIS WILL BE PLACED ON THE USE OF MICROPROCESSORS.

The last sentence was capitalized in the original call for papers. Obviously, the use of microprocessors was the "hot" subject that year. Authors were thus given clear instructions before they submitted a paper: automation of manufacturing is a good subject, but this year's audience is interested in microprocessors. If a submitted paper ignores this special interest, it will either be turned down or, if accepted, might not lift an eyebrow in the audience *regardless of otherwise worthy results shown by the author.*

For company reports, the permissible subject areas tend to be far broader and the requirements less specific. But if you plan to write a report, for either internal or external use, ask your supervisor (and also your colleagues) about the aspect that will be of most interest. Then compile an appropriate circulation list. Identifying and analyzing this readership[37] will require a bit of persistent probing on your part, but it should be an initial aim of your writing objective, because it will help you set your sights on your real purposes.

The master's or doctoral thesis, on the other hand, is primarily directed toward a specific group of readers: your thesis advisor and the thesis evaluation committee. The advisor can give you ample information about the kind of manuscript these readers expect and about their criteria for excellence. If journal publication is also a requirement, you should definitely plan, before writing the thesis, on submitting it or a condensed version to an appropriate journal. Such a plan is discussed in Chapter 17.

After pursuing any of these methods of identifying the interests of readers, you can begin to write your manuscript in a way that will match their objectives with yours.

Matching the Manuscript with the Readership

Now that you have an understanding of your objectives and motivations, and at least a preliminary view of your readers' needs and interests, you are ready to form a strategy for constructing a useful manuscript. The character of your paper will be influenced by your ability to direct information toward your readers.

There are several ways of creating a strong flow of information. These are discussed in subsequent chapters but can be summarized here as follows.

- *Make a plan* (Chapter 4). After you have defined the objectives of your work, assess your potential readers, and select at the start a suitable vehicle for your manuscript—journal, report, conference paper, etc.
- *Orient the reader* (Chapter 7). State your engineering problem or purpose early in the manuscript, and show how it relates to previous work and to current problems.
- *Choose the proper amount of detail* (Chapter 8). Your choices, of course, depend on your sensitivity to the needs of a particular audience. The decision is an important test of your judgment. Too little detail is disappointing and frustrating; too much detail is boring and confusing.
- *Show the overall significance* (Chapter 9). Explain how your paper fills an existing need, how it clarifies a current problem, how it offers a useful application—and why it is important.
- *Emphasize the strong points* (Chapter 10). If you have developed something novel, define what is new and how it is different. Or emphasize the beauty of the engineering design, the utility of the results, the in-

genuity of the method—or whatever is the main contribution in your paper.
- *Get peer reviews.* Be certain to show the draft to technical experts in your organization or at least to knowledgeable associates. They are your test sampling of reader reaction.

These six methods of increasing information flow will help to match the objectives of your manuscript with those of the readership. If the readers subsequently obtain a good match, your paper becomes a useful and satisfying achievement.

Chapter 3

How to Decide Questions
of Multiple Authorship

Who Should Co-Author Your Paper?

The question of who should properly be co-authors of your engineering manuscript is a delicate one for two reasons. First, there is the ethical matter of sharing the credit for the engineering accomplishment and for the paper, as pointed out by Sherrington and Orr.[53] Second, the choice of co-authors will inevitably affect the quality of the paper—not only its style but also its structure and content.

Acquiring co-authors adds a new dimension to planning and writing your manuscript. This cooperative effort can add strength and substance to the paper, but it can also add complications. The ideal authorship team will add the most to the stature of the paper and will experience the least conflict among the contributors. Choosing the best team deserves careful deliberation.

Until about 50 years ago, engineering papers were usually written by a single author. In recent years, because of the nature of modern engineering organizations, a development project has become a complex team effort.[5] Contributors to the completed work may include the project leader, a theoretician, a design engineer, a statistician, a package designer, a system test engineer, and other specialists. Of these, two or more might become co-authors of an engineering paper or report. In the eyes of the engineering community, these authors then become identified with the work. They are the ones who assume the responsibility for recording the character and results of the engineering project. They also bear the public responsibility for its deficiencies or inadequacies.

There are several different ways to choose co-authors, depending on the motivations for preparing the manuscript. You might make your own decision to write a paper or report and then feel that others who had a significant role in your project should help you document it. Or your

14

project leader or supervisor might require that the results be documented, appoint you as senior author, and suggest others to share in the authorship for various reasons, such as

- *Importance of an individual's contribution* (an associate who played a key role in solving the engineering problem)
- *Writing ability* (an engineer who writes well and can help to produce a professionally styled manuscript)
- *Availability to write* (an engineer in your department who has finished his present assignment and can now devote time to writing)
- *Helpful support* (an engineer or technician who provided support work)
- *Prestigious position* (a senior engineer in your organization who is known in this field, who was consulted on the project work, and whose name will add credibility to your manuscript)

Frequently, initial choices are confined to those who made the main contributions to the success of your project. After the writing of the manuscript has started, however, the needs of your department may change. Even with the best of planning, a co-author may have to abandon the writing job because of the press of a new work assignment. Another may take over the work on the manuscript for any of the reasons listed above.

Regardless of your co-authors' qualifications, they will not necessarily be co-writers. You may write the manuscript and then consult with the co-authors on appropriate revisions.

If you parcel out the writing, say one section per author, you have the advantage of concentrated individual inputs. But unless you exercise control over the content and the separate writing styles, the paper can get out of kilter. For example, a long-winded co-author could, because of his writing habits, inflate the lesser portions of the work and thus exaggerate their importance in the manuscript. An overzealous author, impressed with the importance of his or her own contribution, could distort the structure of your paper. Sharing the writing load, then, can be a mixed blessing. The best approach to this question is for all co-authors to agree that the senior author will make most decisions on the structure of the manuscript.

Some of these problems can be resolved by using an anonymous co-author—a ghost writer. A professional writer is likely to be satisfactory when the article or report is a nontechnical version for the general public. When the article is intended for a peer audience, it is usually better for the engineer to write it. If the writing skills of the author(s) are not adequate, the manuscript must be revised by a competent technical editor.

Whether or not all the co-authors write portions of the manuscript, readers will generally associate all their names with the work.

Who's on First?

When a paper or report has several co-authors, the order of names in the by-line is important. The sensitivities of authors run high, but standards for by-lines run very low—so low as to be almost nonexistent. The first name, of course, is the most prominent. Most readers assume that the first name is that of the main author and that the others are secondary. Fashionable trends in the order of names, however, have varied somewhat over the years.

For example, consider the hypothetical case of G. E. White, who designed a new diesel engine 75 years ago. White might have written a journal paper as sole author with the following title and by-line:

The Design of the Five-Cycle Diesel Engine

G. E. White

As an alternative, assume that White asked two colleagues to be co-authors: one was Professor F. R. Horn, his consultant (known for basic work on diesel engines), and the other was I. B. Aaron, White's engineering-design assistant. According to an old tradition in scientific and engineering papers, the authors could be named in the following order:

F. R. Horn, G. E. White, and I. B. Aaron

Here, Horn is the ranking author, but not necessarily the main contributor. This paper would sometimes be referred to in the text of other papers in a truncated form, e.g., as "the design developed by Horn et al."

If the same paper were published today, a modern convention might be followed instead:

G. E. White, F. R. Horn, and I. B. Aaron

This sequence gives precedence to White, the chief designer, and implies that both the design and the paper are mostly his property. This paper would be referred to as one by "White et al."

Another arrangement is used when the ranking author prefers, in modesty, to have his name placed last:

G. E. White, I. B. Aaron, and F. R. Horn

. . . Noblesse oblige.

A different sequence is sometimes used in modern papers. Those authors who cannot agree among themselves about a sensible order, or who wish to avoid the issue, resort to alphabetic order:

I. B. Aaron, F. R. Horn, and G. E. White

But this can result in the incongruous situation in which the assistant Aaron becomes prominent and the primary author finds himself named last. Such a paper will often be referred to as "the design by Aaron et al." because the cart was placed before the horse.

The one saving grace for this kind of confusion is a device used by some in which separate contributions are carefully spelled out in the text:

One of us, G.E.W., originated the basic design concept, and I.B.A. constructed the pilot models.

Probably the best solution to the question of ordering the names is to arrange them according to the relative importance of the authors' work on the project. The principal investigator on an engineering team is an obvious choice for the first name. But when senior professionals of equal rank are to write a paper about their project work, the choice of a principal author is never easy. Alphabetic order may then be appropriate.

A much more difficult problem is the decision on which co-workers to include as authors and which to mention instead in an acknowledgments section. This, too, must be a judgment based on relative importance.

Alternatives to Co-Authorship

Some of your colleagues on the project may feel unfairly treated when they are not selected as co-authors of your paper. Their feelings may be justified if they have made worthwhile contributions to the team effort. There is, however, a reasonable upper limit to the number of authors for a single paper. Some journal editors feel that a paper with more than four or five authors is a bit overburdened. A string of co-authors gives the impression of diffuse achievements, distributed among many contributors. Having too many authors thus defeats the intended purpose. Instead of giving credit where credit is due, it tends to dilute the professional recognition of all.

Potential authors who are not included may develop hurt feelings but may not have even thought of writing their own individual papers. Preparation of separate papers is worth considering if they have made contributions of interest.

Refer to our example of the five-cycle diesel engine. Besides White (the automotive engineer), Horn (the consultant), and Aaron (the design assistant), there were other specialists: a lubrication engineer, a metallurgist, a mechanical engineer, a statistician, and a physicist.

In addition to White's basic design paper, submitted to the *Journal of Automotive Engineering,* the colleagues may find that they can make their own contributions to the literature:

A New Colloidal Lubricant for Journal Bearings

J. A. Herring

(for *Lubrication Engineering,* a trade magazine),

Powder Metallurgical Fabrication of High-Pressure Fuel Nozzles

R. W. Jones

(for *Metallurgical Transactions,* a professional society journal),

Design of a Tapered Piston for Diesel Engines

W. W. Merlin

(for presentation at the 17th Annual Conference of the Society for Automotive Engineers),

Operating Characteristics of a Modified Medawar Fuel Pump

G. L. Harris

(for an internal technical report),

Chemisorption of Volatile Hydrocarbons on Piston Walls

S. T. Bacon

(for *Surface Science,* a physics journal),

An Accelerated Life Test Technique for Diesel Engines

B. R. Mehta

(for a master's thesis and for subsequent publication in *Technometrics,* a journal of statistical methods).

What might have been one paper written by nine authors has developed into a publication program of seven papers, with recognition for everyone on the project. Thus, an engineer's decision to write a separate paper can be a sensible plan provided, of course, that it is not a redundant manuscript.

Summary: The Choices and Decisions

When you decide on multiple authorship, you must make choices that do not concern a sole author. Even with the most harmonious of engineering teams, a joint writing enterprise can have its complications. These may be summarized in the following way:

- *Writing styles* of the contributors may differ.
- *Opposing judgments* can arise about the proportions of the manuscript, even with prior planning.
- *Available time* to work on the paper could be an important variable among the authors.
- *Conflicts among personalities* might have to be resolved.

Even if none of these problems ever arises among your co-authors, you still have to make some important choices:

- *Who will do the writing?* All authors can share the load, each writing a portion. Or the principal author, having the best insights about the engineering results, may write the paper and accept advice on the details from his co-workers. Or the author who has the best writing skills may prepare the manuscript.
- *Who will the co-authors be?* The logical choice is those who have contributed most to the success of the engineering project, especially those who have solved the technical problems. Lesser contributors are mentioned in the acknowledgments.
- *What is the best order of names in the by-line?* The preferred way is to name authors in descending order of their relative contributions. A weak alternative is to use alphabetic order.

Engineers sometimes find themselves appointed as co-authors when the manuscript is half finished or even—as an afterthought—after it is completed. The best way, of course, is to decide the question of authorship first and then to proceed with the planning, as discussed in the next chapter.

Chapter **4**

How to Plan and Organize
Your Paper or Report

The Need for Planning

The engineer's talent for analyzing and organizing can be a great asset
in the preparation of a paper or report. An author can apply the discipline
of engineering planning to the process of writing.

A conventional way to plan a manuscript is to construct an outline.
Some authors, however, claim that they cannot work with outlines—that
instead they can simply begin to write, letting the structure develop as the
writing proceeds. An engineer may feel that planning and outlining are
fine for a student exercise, but that a professional who has already spent
months or years on a technical project does not need an outline to help
organize his or her thinking. In practice, nevertheless, very few engineering
authors can produce a good paper without first drawing up a plan.

Indeed, organizing a manuscript makes demands on the engineer that
differ from those faced in organizing the work project. To the extent that
in both cases an engineering problem is defined and subsequently solved,
the project work and the manuscript go hand-in-hand. But planning a
manuscript also involves additional elements: the interests of readers and
the methods of presenting data. Unless these two elements are considered
in the planning stage, a manuscript about a fine piece of engineering work
can be a dismal failure. Such papers are the bane of journal editors and
their referees; such technical reports miss their mark within an engineering
organization; and such oral presentations fall on deaf ears when presented
at a conference.

There are other reasons for planning and outlining your manuscript.
The first is your own state of flux. Your initial ideas about the manuscript-
to-be will undoubtedly change, and it is far easier to alter drafts of an
outline than to rearrange and rewrite sections of a manuscript. This is true
even if you are working at a word-processing terminal where changes or

Table 1   Objectives of an engineering project

Objectives	Requirements for manuscript plan
To design a new structural form	Features of conventional structure Need for an improved design *Emphasis* on design technique
To show practical applications of known principles	Review of principles to be applied Development of the application *Emphasis* on utility and practicality
To develop a new theory	Statement of the engineering phenomenon Proof of how the theory supports the facts *Emphasis* on development of the theory to explain the phenomenon

insertions and reformatting are done quickly. Manipulating topics on a one-page outline is much simpler and faster than reading and manipulating paragraphs in many pages of text.

A second reason for advance planning applies if you have co-authors. An organizing session is essential, of course, when two or more authors are to coordinate their efforts. Because no two authors would plan a manuscript in quite the same way, it is worthwhile next to consider the elements of a good plan.

Criteria for a Plan

To draw up an effective plan, you should first distinguish between the two kinds of purposes discussed in Chapter 2: the *objectives* of the engineering project and the *motivations* for writing the paper. The objectives are a guide for the technical content of your manuscript; the motivations help you decide how to treat the subject for your readers. Thus, your plan for an engineering manuscript should not be shaped by any standard literary form but rather by your own purposes.

Table 1 presents a few examples that show how various engineering objectives (taken from the list in Chapter 2) require a paper with different types of technical content and distinctions in emphasis. Similarly, your particular motivation will affect your treatment of the subject, as shown in Table 2. After you have given careful thought to those basic purposes—the objectives and the motivations—you have laid the groundwork of a plan for your manuscript.

Ways to Organize

Using cold logic and step-by-step reasoning is not always the best way to organize your paper or report for maximum effectiveness. The

Table 2   Motivations for writing a paper

Motivations	Requirements for manuscript plan
To be published in a prestigious journal	Subject treatment appropriate for the journal chosen Length limited by journal requirements *Emphasis* on the criteria of the editor
To record your developments for their archival value	Rationale for the work project Details of your developments *Emphasis* on methods and results to avoid needless duplication by others
To propose a new program	Problems and needs of the engineering organization Program to meet those needs *Emphasis* on advantages and benefits of proposed program

intuitive approach might be better. Because there is no one best way to organize all engineering manuscripts, the role of the imagination cannot be overemphasized.

For example, writing progress seldom follows the same sequence as progress on an engineering project. Designing a device or developing a process may get off to a false start, or may be sidetracked into a wrong approach, or may undergo modifications before the work is completed. A manuscript describing all these stages of the design or process would be difficult to read. After the problems have been solved in the laboratory, it is time for a new exercise of the imagination: the design of the manuscript. Here, as in the laboratory, is the opportunity for design decisions based on intuitive judgment. John Masefield[36] understood the role of intuition when he wrote: "Man's body is faulty, his mind untrustworthy, but his imagination has made him remarkable."

To construct a strong manuscript, even though you use a standard organization such as introduction-body-conclusion, you need an imaginative approach. Your resourcefulness is brought into play when you build into the manuscript the relationships among

- The engineering information
- The reader's interests
- Your own purposes, i.e., engineering objectives and personal motivations

You can use these relationships as a guide when you gather and organize information for your paper.

The standard methods are as follows. The sources of data are usually your engineering notebooks and progress reports. These should be supplemented with at least a brief literature search to establish what others have done and published in your field. If you work in an organization, the next step is to confer with your supervisor or manager about the kinds of information that should be included. Then the usual procedure is to consult with your colleagues and, if you have co-authors, to plan jointly with them.

In organizing your information consider the following:

- What are the key ideas for the paper?
- Which are the supporting ideas to be subordinated?
- What details need to be included?
- What is the emphasis: the data, the method, your recommendations, a new application, a unique design, etc?
- How long should the manuscript be?
- What should be chosen for the main illustrations and tables of data?
- What information should be relegated to an appendix?

While assembling these preliminary ideas, you can also develop a feeling for the best way to organize by following the suggestions in Chapter 2, "How to Match Your Objectives with Reader Interests."

In organizing information most engineers use outlines of one kind or another. Some prepare an initial abstract instead. I prefer a well-prepared outline, which shows the intended structure and emphasis and clears the air for your writing project.

The Use of Outlines

An outline has two interacting purposes. One is to shape the technical information in logical order. The other is to help organize your thinking. As you gradually construct an outline, it should oscillate between these two purposes until it settles down to an idealized compromise.

Thus, the best way to work with an outline is to keep it flexible. The worst way is to consider it a rigid form into which you must fit your ideas and results.

The two types most frequently used are the topic outline and the sentence outline. I consider the former much easier to work with because it can be read at a glance and can be readily altered. The sentence outline permits fuller development and provides theme sentences for sections and paragraphs, but most engineers prefer the topic outline for ease of use.

The only standard portions of the outline are the introduction section and, depending on content, a conclusions or recommendations section. The choices for the intervening sections are quite important because they represent the author's decisions on the order of presentation of the material, its relative importance, and the level of detail needed.

A representative topic outline for a hypothetical engineering manuscript is shown in the box on the next page. In constructing such an outline, a good method is to set down on a piece of paper the titles of the main sections, leaving spaces beneath each. In our example these are Sections I through V. Your decisions on details and emphasis are then made when you fill in subheadings and sub-subheadings.

Two aspects of constructing an outline are the most critical: How will it relate to your objectives and motivations, and how will it fit into the framework of your readers' interests? The flexibility of outlining can help you adapt your plan to your purposes in the following ways:

- *For a journal paper* the outline shown here is suitably proportioned. Such a structure is generally acceptable to editors and their referees.
- *For an internal report* you would probably expand the plans for Sections III and IV to give reasons for the design choices, the methods that failed, and details of the operating characteristics. These added data might be too lengthy and cumbersome for a journal paper of reasonable length.
- *For a conference presentation* you would offer fewer details than for a journal paper. Slide illustrations tend to be the basis for an oral presentation, and the materials that can be placed in simplified charts and diagrams are your logical choices.
- *For a trade magazine* the theoretical section is best minimized or omitted. Sections III and IV on design methods and performance data will be of most interest. The conclusions section might be devoted to a summary of efficiencies and cost savings, or it could be omitted. Charts and illustrations may be less detailed than for a journal paper.
- *For a proposal* you would need a greatly expanded introduction, probably preceded by a short section summarizing the report for the executive reader. Sections on laboratory facilities and personnel qualifications are essential. There would probably be no section on operating characteristics. Potential advantages are heavily emphasized, as is the ability to fill the customer's needs.
- *For a thesis* the introductory section needs to be enlarged to include a historical review, a detailed literature search of previous work, and a discussion of the rationale for the design approach. Sections III and IV might be replaced by an "Experimental Results" section giving data on a laboratory version of the collector array. Sections V.A and V.B on the significance and the thermal efficiency should be lengthier and more detailed than in a typical journal paper. The appendices are usually more voluminous than most journal editors would accept for publication.

After you have established your purposes, thought about your readers, and constructed a preliminary outline, review it to see whether its content

DESIGN OF A HIGH-EFFICIENCY SOLAR ENERGY SYSTEM

B. A. Smith

Energy Technologies, Inc., Wayward, Massachusetts

ABSTRACT

I. INTRODUCTION
  A. Current design trends in solar energy
  B. Need for higher efficiencies
  C. New design approaches
    1. Dual-concentrator collector array
    2. Improved transport fluids

II. THEORY OF THE DUAL-CONCENTRATOR COLLECTOR ARRAY
  A. Principle of multistage concentration of light rays
    1. Use of both Fresnel lenses and mirrors for tracking arrays
    2. Method of successive concentrations
  B. Comparative analysis of collector arrays
    1. Energy gain calculations
    2. Effects of adding successive array levels
    3. Parasitic heat losses
  C. Theoretical limits of energy concentrators
    1. Comparison of lens and mirror configurations
    2. Choice of optimum design

III. APPLICATION TO SOLAR ENERGY DESIGN
  A. Basic design of the collector
    1. Lens system
    2. Mirror system
    3. Minimization of heat loss
  B. Design of storage tanks, circulation loop, and heat exchanger
  C. Selection of transport fluid
    1. Properties: high heat capacity, low corrosion
    2. Advantages over water

IV. OPERATING CHARACTERISTICS
  A. Collector array performance
    1. Incident solar energy
    2. Collector efficiencies
  B. Heat transfer performance
    1. Using water as transport fluid
    2. Choice of 64/36 ratio of propylene glycol/water
  C. Cost efficiency
    1. Data for dual-concentrator design
    2. Comparison with flat-plate array designs

V. CONCLUSIONS
  A. Significance of the concepts
    1. Multistage concentrator
    2. Transport fluid with high heat capacity
  B. Energy efficiencies of the system
  C. Limitations of the design
  D. Cost savings
  E. Potential application to other design problems

VI. REFERENCES

VII. APPENDIX: Derivation of energy gain algorithm

YOUR READERSHIP

	Has need for new energy sources	Is interested in improved methods	Looks for cost savings	Requires more efficient use	Seeks more background information
Develop a new theory		I.C II.A		I.B II.B	I.A
Show applications	I.A V.E.	III.A III.B V.D	IV.C	III.C	VI
Give characteristics		IV.A	V.D	IV.C V.B	

(left margin label: YOUR OBJECTIVES)

Figure 1   A matrix analysis of the outline for "Design of a High-Efficiency Solar Energy System."

matches the needs of your anticipated audience. One way to review the outline is to analyze it with the aid of a matrix. Figure 1 is an example of how to draw up such a matrix. Filling in the cells with alphanumeric codes for the outline headings shows how the objectives are related to reader interests. The existence of blank cells can suggest items to be added to the outline. Thus, your future manuscript, besides being logically organized, will also fill a need for its readers.

You are now ready to write the manuscript. If you are engaged in design or development work and have not yet finished the project, you can still start the manuscript, as suggested in the next chapter.

Chapter 5

How to Use the Incremental Method

The Conventional Method

The time-honored sequence in engineering achievements is to complete the work, write a paper, and get it published. But some engineers who attempt to write after a development project is finished face an obstacle course replete with roadblocks. The path to a finished manuscript can be a tortuous one for several reasons.

The first of these is the nature of engineering organizations and the proliferation of work projects. After a development program is completed, any talented engineer tends to be in demand for the next assignment. The press of new work may have a much higher priority than writing the intended paper, and the job of writing about the project just finished may have to be relegated to a colleague who has less writing skill and perhaps less insight into the nature of the engineering results. The delays in finally getting started on the manuscript can extend for months or years—long after anyone will be interested in writing it or reading it. But with adequate planning and adequate personal motivation, an author will record the achievements promptly and share them with the peer community.

There are, of course, additional obstacles to writing. Even if generous time is allotted for preparing a manuscript, an author's attitude and the work environment can be stumbling blocks. For example, the engineer may have been working on a technical problem for an extended period. The logical reasoning and the novel techniques that had seemed so fresh and exciting during the course of the program now seem faded and stale. The loss of perspective is stultifying. In writing the paper, the engineer is now using the same timeworn phrases and stylized expressions used previously in progress reports and in presentations to management. Boredom is now reflected in the document—a tired, lifeless manuscript lacking in

27

imagination and failing to bring out the ingenuity and beauty of the author's laboratory accomplishments. This boredom can be eliminated from such a paper by rewriting and perhaps by extensive editing. In addition, peer reviews will suggest how the author can revise further to clarify the nature of the technical contributions.

Moreover, if the author works for an engineering company or government agency, the manuscript is subject to the further delays of management review and approval for publication. Quite a bit of time can elapse between the termination of the work project and the day the manuscript is ready to be submitted to a journal.

Subsequently, it will take a few months or perhaps a year for such a manuscript to be reviewed by the journal editor, sent to referees for appraisal, possibly revised once more, prepared for the printer, and finally published. It is not surprising that many papers in engineering journals appear long after the author's laboratory work was completed.

With experience, some authors learn to minimize such delays, but for others the delays are a continual frustration. Anyone who has developed a new design, process, material, or engineering technique deserves to be known among his peers as a contributor to the state of the art. Besides, getting the paper into print *promptly* is important because of the competitive nature of engineering developments.

Many identical innovations—patentable or not—have been developed simultaneously and independently. A famous example is the invention of the telephone, for which Alexander Graham Bell and Elisha Gray filed separately for a patent on the same day. Another is the process for making street gas by passing steam over red-hot coal. This scheme was devised independently in France by Tessie du Motay and in America by Thaddeus Lowe. One more example is the method of making steel by blowing air through molten iron to burn out the carbon. This famous process was developed simultaneously by William Kelley in America and by Sir Henry Bessemer in England. You might not invent anything as earth-shaking as the telephone or the Bessemer furnace, but if you develop a faster electronic circuit, a stronger synthetic fiber, or a more lethal mousetrap *you should publish as soon as possible,* especially if you decide not to apply for a patent.

If the conventional method of writing and publishing a paper is not fast enough for you, try a different approach, which I call "the incremental method." In this procedure you write the manuscript in successive segments, starting before the project work is completed. Although this approach might not seem feasible at first, it has been used with success by many engineer authors, and you should consider the techniques outlined here.

Writing in Increments

The time to start your manuscript is at the beginning of the development program. From the plans for your project, you can make your first estimate of the content of your forthcoming paper. A plan for formal writing then becomes part of the work itself and, in a subtle way, a guide for this work. The insights you develop while writing can strengthen the thread of engineering developments and add to the accomplishments on the project, as I suggested some time ago in a paper on creativity.[41]

The first step toward writing in increments is to construct an outline, as mentioned in Chapter 4. The details of the outline, and perhaps even the main headings, may change in future months, but no matter. You have made a start. In your role of prospective author you have given yourself a psychological boost.

Early in the project, as indicated in Figure 2, you can write a brief, tentative abstract based on your engineering goals. Before long you will have at hand the kinds of information that appear in a good introduction: the purpose of the work, the relation to previous papers in this field, a statement of the novelty of the engineering approach, and a short preview of what the manuscript is to contain.

In the ensuing weeks or months, you can devote short periods of time to writing successive sections of the manuscript, building up the content in manageable increments. These portions should be typed in draft form and set aside. Each section can be devoted to one of the several themes in your paper and may be a self-contained segment, as suggested by Tracey.[60]

For a specific example, refer to the outline in Chapter 4 on "Design of a High-Efficiency Solar Energy System." As an engineer on this project, your first step undoubtedly would have been to develop the theory of the collector array, the topic of Section II. Before you began to apply your theory to solar energy design, you would probably have been able to write most of Section II. Further inspection of the outline shows that as the work proceeded you could develop corresponding portions of the manuscript, as implied in Figure 2.[44, 46]

Near the completion of the project, you would be ready to write the conclusions section. Thus, after the development program was finished, you would not be in the customary position of planning to write it up for publication. Instead, the accumulated sections of your paper would be ready for revising, polishing, and final approval. The paper would then be ready to submit to a journal as a current and timely piece of work.

The principle of writing journal papers in increments has several important advantages over the conventional method. The incremental procedure makes writing an interesting and productive part of your work

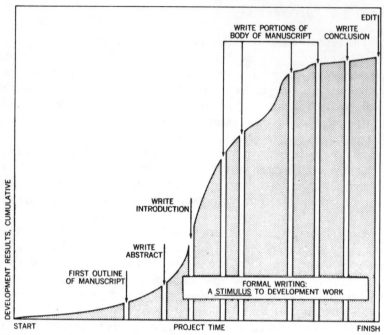

Figure 2 Writing a manuscript in increments during an engineering development project. (From reference 44.)

instead of the traditional task of recording results. Writing your manuscript for example, may suddenly reveal to you the necessity of taking more data—while the equipment is still set up. The discipline of writing thus combines with the discipline of engineering effort in unexpected ways, and after the manuscript is written you will seldom remember which actually came first—the chicken or the egg.

There is a strong but subtle interaction among author, work, and manuscript, from which all three stand to benefit. The author experiences the psychological advantage of early progress on the paper. The development work reflects the fresh insight and new understanding that normally evolve from the writing. And the paper, being conceived and born during the excitement of accomplishment, has become easy to write and interesting to read.

Chapter 6

How to Write an Abstract

Types of Abstracts

For an engineer the abstract of a paper or report serves one of two special purposes. For the *engineering reader* it reveals in a nutshell the main findings and significance. For the *engineering author*, preparing the abstract requires a statement of the central idea in brief, unmistakable terms, a beneficial exercise for any writer. When properly written, the abstract becomes a definitive piece of writing for both author and reader, but when it is poorly written, *both* are at a disadvantage.

Like any other kind of writing, an abstract with weak structure will not serve its purpose. The form should be chosen to fit the function. There are three types of abstract, each having a somewhat different purpose:

- The *indicative*, or descriptive, type states the general subject matter of the document that follows. It tells in a qualitative way what the report contains.
- The *informative*, or informational, abstract highlights the findings and results, briefly but quantitatively. It is a condensed version of the engineering work, without discussion or interpretation.
- The *informative-indicative* abstract is a combined form that gives specific information about the principal findings and results and general information about the rest of the document.

The type and length that is best for your manuscript depends on the publisher, as discussed in detail by McGirr.[38] If the vehicle is to be a journal, consult the editor's style guide and be guided by the author abstracts in recent issues of the journal. If the vehicle is to be an engineering seminar, look at the proceedings from the preceding year.

31

A decision on the type of abstract also depends on the subject matter. The reader of a review paper or a survey generally needs a "road map," which is best provided by a descriptive abstract. An experimental or theoretical study should usually begin with an informative abstract that gives appropriate details. The abstract in a student's thesis can be either an informative or an informative-indicative type, in which the final result is mentioned specifically and the various kinds of background and supporting information are mentioned only in a general way. Examples of several kinds of abstract are given at the end of this chapter.

No matter what type of abstract you choose, rest assured that it will be more widely read than the document itself. Many readers will not read your full paper because the abstract may give them all the information they need or because your general topic seems of no immediate interest. For these reasons even a short abstract must include all the pertinent elements of information.

The Required Information Elements

You should be especially sensitive to the needs of your audience when you write the abstract. Use only essential information. As already pointed out, a much larger number of people will read your abstract than your full paper, and many will have broad interests. You need to orient them to your subject in an efficient manner, using great economy of words. It is, of course, far easier to write a long-winded report, full of helpful explanations, than it is to write a few terse, efficient sentences without an irrelevant phrase or clause. A lengthy abstract defeats its purpose. In general, 200 words is a sensible maximum for a relatively long paper or report; 50 to 100 words may suffice for a short paper.

Although your abstract should be brief, usually confined to a single paragraph, it must contain certain elements of information, which I have discussed elsewhere.[43] The entire abstract can consist of as few as three sentences. An informative or informative-indicative abstract should contain the following three elements:

- *A statement of the engineering problem.* Boil this down to a single sentence. This kind of writing clears the air of any hazy expression of your real purpose.
- *An explanation of your approach to solving the problem.* The "approach" might be an analytic method, a design technique, a system concept, a device improvement, a testing scheme, etc.
- *The principal result.* Your net accomplishment may be a theoretical or an experimental finding, a new design with improved properties, a recommended course of action, etc.

Writing concentrated information is not a trivial job, one of composing a few summarizing sentences. Rather, it is a hard discipline, a test of your sense of relative importance. To write a good abstract, you must strip away the peripheral, supporting information in your paper and expose the hard core of your contribution, defining the real substance of your manuscript in the simplest terms. Your abstract is then a subminiature version of the body of your paper, self-contained and unambiguous.

To write so concisely, you must omit certain elements of information. Do not use figures, tables, or literature references. Exclude from the abstract any information that is not based on the content of the paper.

In addition, there are other elements that conventionally appear in other sections of the manuscript but do not belong in the abstract. For example, a brief historical review, with reference to previous work in the field, is suitable for the introduction. A discussion of the significance of an internal report is generally relegated to a separate summary for the executive reader, if such an initial section is required by the engineering organization. An evaluation of a new technique or a discussion of the validity of the results should properly appear in a conclusions section.

Up to this point, the abstract has been considered an adjunct to your manuscript, providing a bird's-eye view of the sections that follow it. Your abstract, of course, is also used for other purposes and in other contexts. It may be recorded, sometimes verbatim, by various abstracting services that publish reference volumes used for literature searching. These include *Engineering Index*, *Chemical Abstracts*, *Science Abstracts*, and many others in more specialized fields of engineering, such as *Computer Abstracts*, *Metals Abstracts, Control Abstracts*, etc. There are also a number of computerized searching services, such as Lockheed's *DIALOG*, System Development Corporation's *ORBIT*, and Bibliographic Retrieval Service's *BRS SEARCH*. These maintain extensive data banks obtained from the numerous abstracting services. The commercial information retrieval systems provide computer printouts of bibliographies, with abstracts, on any technical subject requested. (See Chapter 16.)

The automated literature search, provided either in-house by large companies and government facilities or externally by vendors, is becoming widely used by engineers. A compilation of abstracts helps the requester to determine from an abstract whether the full paper or report would have the needed information. If so, the engineer can order a copy of the complete document.

A good choice of information elements thus makes your abstract a potent and influential piece of "microliterature" in two kinds of situations. First, it frames your ideas for those who are beginning to read your paper. Second, it identifies your contribution for those who are using an information-retrieval service. When your choice of information elements is poor, you discourage both classes of readers from looking at your full paper.

Your efforts will then be lost to part of the engineering community. In the huge volume of modern engineering literature your manuscript, too, can then be "lost."

The examples that follow show how the selection of information elements can radically affect the quality of your abstract.

Illustrative Examples

A convenient way to show the nature of a good abstract is to use as a basis the outline in Chapter 4 for our hypothetical paper by Smith, "The Design of a High-Efficiency Solar Energy System." A *descriptive abstract* with the three essential elements can be illustrated as follows:

> The successful design of cost-effective solar energy systems depends mainly on the development of higher conversion efficiencies. A theory is given for the thermal performance of a new solar collector array that combines reflected and refracted light rays. The theory is applied to the design of a complete system for solar heating of a building. Dynamic operating characteristics for a one-month interval are given for the collector array and heat transfer devices, and cost efficiency is compared with that of conventional designs.

This descriptive abstract has several desirable qualities: it is short and self-contained; it provides key information elements; and it serves as a miniature table of contents. An engineer scanning a journal can tell at a glance from this abstract whether it will be worthwhile to read the full paper.

Moreover, this kind of information, although purely qualitative, is useful for an information-retrieval system that deals only with abstracts of published papers and reports. Some of these systems automatically search for key words previously ferreted out by an information specialist. In the case of Smith's paper, the specialist picks out the following key terms from the abstract: *solar energy, solar collector array, solar system, solar energy system design, solar heating, heat transfer,* and *cost efficiency.* The paper is then entered into the data bank by author and title, along with these key words. Subsequently, anyone requesting from the system a bibliography on solar heating, etc., receives a group of abstracts, including the one for Smith's paper.

Another type of retrieval system does not search for *assigned* key words, but instead automatically searches the full text of the abstract. In this case, the computer retrieves Smith's abstract from the data bank because of certain specified words that happen to appear in the author's abstract. Therefore, the text of a good abstract should contain key words.

A poorly written abstract suffers from ambiguity and omission of key elements. Here is how such a descriptive abstract might be written for the Smith paper:

Modern solar systems for converting the sun's energy into heat are gradually coming into wider use, but the efficiency of various system components has inhibited the growth of the industry. This paper is a study of a dual method for concentrating the sun's rays and of removing more heat from the circulating medium. The efficiency of the design is shown to be higher than that of the Wooster system.[1] The design improvements show new potential for solar energy applications.

The writer of this paragraph has little empathy for his readers. The first sentence is a brave attempt to set the stage for an explanation of the development work. The sentence, however, is gratuitous for any engineer who is sufficiently knowledgeable about solar energy to be at all interested in system design. The second sentence is a misguided attempt to be clear by mentioning the concentration of the sun's rays and the removal of thermal energy from the heat exchanger. More appropriate wording for the engineer reader who works in system design would be the familiar "solar collector" and "heat transfer." These would have been likely key words for a literature search. In the next sentence "efficiency" is ambiguous. To the uninitiated reader, the efficiency might have been determined either theoretically or experimentally, or both. Nor is it clear whether the efficiency is measured by static or dynamic means. Also, the citing of reference 1, the article about the Wooster system, is inappropriate because it violates the principle of quick, complete, easy-to-read information in an abstract. The last sentence is a remark about design improvements that is merely a "sales pitch," revealing nothing of interest. The one virtue of this abstract is its length—only 79 words. It has few other redeeming features.

The *informative abstract*, instead of indicating the general content, should be specific and quantitative, giving only essential data. The following example shows how this type would be written for the Smith paper:

In the design of a solar energy system using arrays of multiple solar panels, the Hottel-Whillier-Bliss (HWB) model is generally used to evaluate steady-state efficiency. The HWB equation did not accurately predict dynamic thermal performance for the present system, which uses a novel collector array. The theoretical limit of energy gain for this new array is 6.5×10^6 BTU, which extends into the nonlinear mode of the HWB equation. A modified form of the equation was, therefore, developed for improved linearity, effective for collector efficiencies up to 54%.

The new collector array, using the principle of successive concentrations, was incorporated into the design of a complete system of solar heating of buildings. The transport fluid for transferring energy from the solar array to the storage tank was important to overall efficiency. An optimum ratio of 64/36 was determined for the proportion of propylene glycol to water. Dynamic performance tests during the month of July 1980 in Phoenix, Arizona, yielded an average efficiency of 49% for a gross collector area of 721 m^2 when the array faced south at an angle of 35° from the horizontal. Losses between collector and storage tank were 3.8 × 10⁸ J per month. The cost efficiency of the system was a 9% improvement over that of the SOLTHERM system operating in Phoenix during the same month.

This example shows how net results can be cited from various portions of a paper to give a clear picture of the significant contents.

The *informative-indicative abstract* offers fewer details, instead giving emphasis to the author's chief contribution. In doing so, the following abstract highlights the same technical information that is emphasized in the complete paper:

The successful design of cost-effective solar energy systems for heating buildings depends chiefly on high-efficiency conversion of light energy into heat. A theory is given for the thermal performance of a new solar collector array. The Hottel-Whillier-Bliss (HWB) model is modified for accurate prediction of the energy gain of the new solar array. This three-stage collector, using the principle of successive concentrations, was incorporated into the design of a complete solar energy system. An optimum ratio was determined for the proportion of propylene glycol to water for the fluid transferring energy from array to storage tank. In dynamic tests during July 1980 in Phoenix, Arizona, the average efficiency was 49% for a gross collector area of 721 m² when the array faced south at 35° from the horizontal. Losses between collector and storage tank were 3.8 × 10⁸ J per month. The cost efficiency compared favorably with that of the SOLTHERM system operating in the same area that month.

In this example the theoretical model and the configuration of the array are mentioned in general terms. For many readers the main interest is how well the new system will perform, and this informative-descriptive abstract fills the need.

In summary, the key to choosing the best type of abstract for your paper or report is your understanding of reader interest. For example, the program committee of an engineering conference will often use the authors' abstracts as the sole basis for accepting papers to be presented. For this purpose, and also for formal reports and theses, the informative abstract is probably best. Most engineering journals with specialized readership prefer the informative-indicative type. A magazine with a broad audience of readers in various fields of engineering may prefer the indicative abstract. In any case, the style requested by the publishing vehicle is the best criterion.

Many engineering societies and government agencies publish instructions for preparing an abstract. In addition, the American National Standards Institute provides an established standard [2] for abstracts, and the recent book by Cremmins [17] is an authoritative treatment of the preparation of abstracts.

Chapter 7

How to Construct the Introduction

First Things First

For most authors the introduction is the most difficult and troublesome section to prepare. Even after methodical thinking, planning, and outlining, it is not always easy to find just the right words and the best way to introduce your subject to a particular set of readers. After drawing up a good outline, you should be well prepared to write about your work and the results, but you may not have a ready answer for the initial question: How much background do your readers need early in the manuscript, i.e., how do you best orient your audience? This is a many-faceted question.

For your more knowledgeable readers a thorough orientation about your project may seem patronizing and inappropriate; for others, less experienced, too little background information will leave them puzzled and uncertain about your purposes.

The answer, then, lies not only in the kind of manuscript you are writing, but also in the type of readership. The examples given in this chapter show how your introduction can be constructed for each of the several types of engineering manuscripts. In each case we use the basic outline on the design of a solar energy system (Chapter 4).

There is one standard rule for writing your introduction: you must consider it to be independent of the abstract and write it as if the latter never existed.

Moreover, in preparing your introduction you must be far more sensitive to reader interest than you were in writing the abstract. You need to include more details about the objectives and the background. You might consider some or all of these conventional information elements:

- The purpose of the manuscript
- A definition of the engineering problem

- The background of previous work in this field, including different approaches
- The chief contributions of others
- The scope of the manuscript
- The rationale for the project
- A brief indication of the technical contents to follow

The flavor and content of the introductory section will depend on the purpose of your manuscript. The introductory part of a company report should concentrate on the project itself and its objectives. In a paper for journal publication the introduction should emphasize the nature of your work in relation to previously published papers. In an article for a trade magazine the introduction usually highlights the engineering problem and how your solution fills a need. In a thesis the introduction tends to give detailed treatment of all the listed information elements.

Illustrative Examples

Internal Report

When your manuscript is intended solely for use within your engineering organization, the opening paragraphs might be slanted toward local interests in the following way:

INTRODUCTION

Early in 1976 it became apparent that the flat-plate collector array could have only limited use in the design of high-efficiency solar energy plants. A new optical design for the collector array,[1] proposed by George Edison of the Advanced Development section of the Solar Laboratory, led to the initiation of the SOLRAY project for solar system development.

Conventional solar energy heating systems attain overall efficiencies up to 10 percent. The design goal of the SOLRAY system was 13.5 percent thermal efficiency based on projected improvements due to the new collector array and to selection of a transport fluid for enhanced heat recovery. A second design objective for the SOLRAY project was to develop a simplified structure for the new tracking collector to reduce its manufacturing cost below that of conventional flat panels.

This report gives the theory of the dual concentration of light radiation and shows the limitation on energy gain. The results are then applied to the design of a complete solar energy system for residential heating, using an optimum design for the collector, a novel array tracking mechanism, and a radically improved heat exchanger.

Data for collector array performance, obtained by the Reliability and Test section, are reported for a dwelling installation in Phoenix, Arizona, in July 1979.

This introduction has the unmistakable flavor of a document for internal consumption.

The next example shows how a manuscript, written for the eyes of the outside world, can have a decidedly different tone and content.

Journal Paper

Although an internal report is sometimes submitted in its original form to an outside journal, it is usually better to modify the manuscript according to the style and needs of a particular publication. In the environment of technical journals, the editor and the readers are more interested in what you have contributed, in the framework of existing literature, than in your internal politics and the housekeeping problems.

Your journal paper might start out as follows:

INTRODUCTION

Solar energy systems for residential use are at present economical only in certain geographical areas that have consistent solar radiation. Current efforts at cost reduction of solar heating systems are generally devoted to the development of low-cost components and to the improvement of thermal efficiency.

Advances in component design have been mostly in the area of collector panel arrays. The pioneering efforts of Heath[1] in the invention of the parabolic trough collector array, and of Sonnenschein[2] in its further development, have shown that a combination of incident and reflected light can increase the efficiency of solar panels by 42 percent. The high initial cost of such collectors, however, has limited their use in practical systems.

This paper gives the theory of the dual-concentrator collector array and reports on its application in the design of a prototype model of a high-efficiency solar heating system. Unlike the parabolic concentrator, the new collector array uses a combination of mirrors and Fresnel lenses to provide successive concentrations of solar energy on the surfaces of the absorber coils. The lens configuration occupies one-half the spatial volume required by the hyperbolic mirror design reported by Perrier[3,4] for successive light concentrations. The energy gain of the new collector reported here was determined from the Holmes modification[5] of the Hottel-Whillier-Bliss equation.[6]

A second contribution to the understanding of system efficiency is an analytic study of transport fluids, which yielded an optimum ratio for obtaining heat transfer from the collector to the storage tank.

As shown recently by Arras,[7] static performance data for solar systems are not directly related to dynamic performance. Therefore a comparative study was made in Phoenix, Arizona, of our prototype model and the older SOL-THERM system, which was described in this journal by Holtzmann.[8] The data give comparisons of cost efficiency, as well as thermal performance.

In this example the author is careful to point out the similarities to prior published work and to indicate the areas of novel development.

The formalities of wording and structure would have to be modified considerably if this paper were to be given orally at a technical symposium. The introductory remarks for the same paper could be styled as in the next example.

Conference Paper (Oral Version)

Most papers given at technical seminars are built around comments on a series of slides, flipcharts, or other visual displays. Some of the details of a journal paper are omitted, and the introduction sets the stage for showing charts, diagrams, tables, or photographs:

INTRODUCTION

Designing solar energy systems today usually involves improvements in the form of the collector arrays and the development of more efficient methods of heat transfer. We have designed a collector array based on a new principle of successive concentrations of solar radiation, using both Fresnel lenses and parabolic mirrors. We have also made studies of the optimum composition of a transport fluid for conducting heat into a storage tank. Our design has now progressed to a prototype model, and we have run some tests of dynamic properties in a model house in Arizona.

Our principle of concentrating the light rays in two successive stages is a modification of the Perrier arrangement of hyperbolic mirrors, which is illustrated here in the first slide.

These brief introductory remarks are confined to what was done and what will be shown. The oral version of a paper usually has many more visual illustrations than are needed for the written version.

Trade Magazine Article

If you decide to write a short article for a technical trade magazine, your introduction should emphasize the engineering techniques that you have developed and the utility of the system. Here is the kind of direct approach that you might adopt:

INTRODUCTION

The configuration of the light collector array is one of the main design problems in modern solar heating systems. The configuration of the collector, together with the minimization of heat loss, contribute more to system efficiency than any other design parameter.

New techniques in these two areas have been incorporated into a prototype model of a complete solar energy system for residential heating. This article shows the design methods for a new collector array that uses simple optics consisting of a mirror and a group of Fresnel lenses to concentrate the light energy in successive stages. A new analytical method also shows how to determine the best ratio of propylene glycol to water for the transport fluid. Included are test data for dynamic performance and cost efficiency.

These brief, direct statements permit you to launch directly into the sections on the design methods and the operating features that are the basic interest in the trade press.

Thesis

In writing the introduction for your master's or doctor's thesis, you begin with a special advantage: you already know your readers and their requirements. Unlike the question of broad readership for reports and papers, there is no doubt about your primary readers for the thesis. They are a select and limited group—your thesis advisor and the thesis evaluation committee at your school. You should know, for example, whether the work must be original or whether a review will be acceptable, how much background material is advisable, whether you need to provide your own experimental data, what sort of emphasis is expected, and what to avoid in the manuscript.

For example we assume that you have been working in the solar energy industry and are obtaining a master's degree in mechanical engineering. Your advisor likes the topic of solar energy system design; he expects you to include your theoretical and experimental results; and he suggests that you review the background of research in collector arrays. Your thesis can get off to a strong start if your introduction mirrors your advisor's general suggestions. Here is how you might write the first section.

INTRODUCTION

The use of solar energy for residential heating has progressed in recent years and has found commercial application in a number of geographic areas. The extension of its practical use depends on further improvement in overall thermal efficiency of solar energy systems. This study discusses the design of a system based on two methods of enhancing the efficiency. The first is the method of converting radiant energy into thermal energy by a novel design of the solar collector array. The second is an analytic method for finding the optimum composition of the transport fluid that carries the heat energy from collector to storage tank.

Historical Review

Concentration of the sun's rays to provide heat has been tried in various ways since ancient times, but the first practical system was developed by Lavoisier[1] in 1787 for scientific experiments. Early concepts of mirrors, lenses, and circulation pumps had no commercial significance until the development of the flat panel array by Jones[2] in 1935. Gradual enhancements of efficiency led to the fabrication of practical solar array panels for thermal energy systems in 1940 by Smith.[3]

The continual quest for better efficiencies concerns not only the choice of panel materials, which are reviewed in a recent exhaustive study by Smart,[4] but also the configuration of collector arrays and paths for heat transfer throughout the system.

There are three basic types of collector arrays:

1. The flat panel, in which a metallic coil filled with a transport fluid rests on a flat insulated bed and absorbs incident sunlight energy. The principle is described by Smith.[3]
2. The Fresnel lens concentrator, which provides higher light intensity for the absorber coils than does the flat panel. The first design was reported by Brown.[5]
3. The parabolic concentrator, in which a curved mirror causes sunlight to converge on the absorber coils. An automatic tracking mechanism continuously turns the panel so that it faces the sun during all daylight hours. This concept is due to Hollister.[6]

These three types are currently evolving into more advanced designs having higher efficiencies than arrays in current use.

Design Approaches

The new collector array concentrates the light rays in two stages, first by means of a curved mirror and subsequently by a bank of Fresnel lenses placed between the mirror and the absorption coils. The optical design of the lenses provides automatic tracking of sunlight without the need for light sensors and servomechanisms to position the panel mechanically.

Methods of measuring the thermal efficiency of the collectors depend on the standard Hottel-Whillier-Bliss equation.[7,8] The complexity of the new collector design required a modification of the equation for accurate determination of energy gain.

The heat transfer problem is approached by a new analytic method, a modification of the Strauss technique[9] for finding an optimum concentration of the transport fluid. This procedure, which determines the best proportion for minimum corrosion and maximum specific heat, was used for choosing the propylene glycol/water mixture.

Home heating installations involve the use of large arrays of collector panels and the installation of corrosion-free piping systems of reasonable cost. The ideal system of the future will use arrays with far higher thermal efficiency and lower initial cost. The research described here is directed toward those three goals and deals with the details of several new design concepts.

The following sections present the theory of successive light concentrations, its application to system design, the operating characteristics of the energy plant, and a critique of the design methodologies.

Because of the scholarly thoroughness expected of a degree candidate, the introduction in many theses is longer than this example. Mere length, however, is not evidence of scholarship. An overly long introduction defeats its purpose and weakens the paper by diluting it with wordy passages and "padding." Your orientation of the reader, however lengthy, should not be watered down with trivia.

For further details on how to prepare theses and dissertations, refer to Chapter 17 and also to the book by Sternberg[55] and to those by Teitelbaum and by Gibaldi and Achtert in the list of Additional References, p. 175.

Chapter 8

How to Prepare the Body
of Your Manuscript

The Main Sections: Your Basic Structure

Unlike the challenge of orienting your *reader* in the abstract and introduction, the challenge of writing the main body of your manuscript is one of orienting the material in the *manuscript*. When you begin to write, you may decide to modify your initial outline as you develop your ideas for the best structure.

Writing, of course, is a dynamic process, subject to false starts and sudden inspirations. At best, the outline is a first approximation, and to follow it slavishly would be a mechanical and unimaginative way to compose your paper. The initial "blocking out" of ideas into the form of an outline is a bit like the art of modeling a clay statue. The artist first blocks out the general form of his composition and then gradually removes or adds clay to bring out the structural detail.

In the same way the insights and refinements that occur to you while writing will bring out the significance of your work. Some of the items that you add to the skeletal frame of the outline are your supporting data: charts, diagrams, photographs, or tables. And, like the sculptor, you find that removing some of the material strengthens the structure rather than weakens it.

Refining the Structure

The structure of the main sections can be built up in ways that strengthen your own concept of the manuscript. The first of these supporting methods is the adoption of good, descriptive headings, i.e., the short titles that define sections and subsections. The choice of these headings is important because they break up your manuscript into manageable

44

portions, easily seen and identified. Well-chosen headings are not only an aid to the reader but also a reminder to the author to keep in focus the content of each section.

For an illustration, refer once more to the outline of our hypothetical paper in Chapter 4. At first, all the section titles I through VII appear to be proper headings for the seven sections of your paper or report. But as you write the manuscript, you begin to feel that some of the titles that seemed so logical and straightforward in the outline might not be the most appropriate after all. The title of Section III, APPLICATION TO SOLAR ENERGY DESIGN, appeared to be a natural sequel to Section II on the theory of the collector array. As you develop Section III, however, you find that III.B (the design of the storage tanks) has turned out to be just as important as III.A (the design of the collector array). As an afterthought you change the heading for Section III to the more comprehensive DESIGN OF A SOLAR ENERGY THERMAL SYSTEM. This turns out to be more descriptive for another reason. There are two kinds of solar energy systems: the thermal type for converting sunlight into heat and the photovoltaic type for converting light into electricity. Your new heading is thus more definitive and is a decided improvement.

Another refinement is the way you use graphics to illustrate your engineering concepts or designs. Drawings or photographs can give strong support to the structure of your paper, but their effectiveness depends, of course, on how they are designed and where they are used. The best graphic aids are those styled to serve one of six definite purposes:

- To describe a function
- To show the external appearance
- To show internal structures (cutaways and exploded drawings)
- To illustrate a phenomenon
- To demonstrate relationships
- To define novelty and originality

Graphic aids, however, may not strengthen your manuscript. If you have not designed a photograph or drawing for its purpose, it may dilute and actually weaken your manuscript. For example, if your report about a mechanical device is intended to show its operational features, a photograph of the external housing will not support your explanations in the text of the device functions. Or if your paper concerns a new technique for physical measurements, graph charts that give the results of the measurements may not show the novelty of the technique itself. The choice of appropriate illustrations and the ways of making them most effective are discussed in detail in Chapter 13.

The table is another useful device for supporting the structure of your paper. Lists of tabular data can clarify, emphasize, and even dramatize information in ways that cannot be done in the body of the text. When properly constructed, the rows and columns of information in a table permit easy comparisons and interpretation of trends. Like photographs and drawings, tables are an asset only if carefully designed. Chapter 14 gives examples of table structure and offers a rationale for proper design.

The Appendix to the Main Body

The appendix provides a much different kind of aid to the structural qualities of your manuscript. By omitting certain kinds of information from your main sections and placing them in the appendix, you can strengthen the structure and improve the readability of the more important portions of your manuscript. The appendix is useful to only a limited part of your readership. It may contain reference data that are too detailed for some of your readers but are welcomed by others. One example is the derivation of mathematical equations; such material can be relegated to the appendix for use by interested readers. Other examples are supplementary tables of data, descriptions of equipment or procedures, and analyses that support your main findings but are not primary contributions.

Another useful feature of the appendix is the flexibility it lends to the length of a paper for publication. Among journal editors, opinions will vary about the value of the material you have placed in the appendix. The particular editor who receives your manuscript may decide to delete all or part of your appendix, which then serves as a convenient safety valve to accommodate the pressure of arbitrary editorial judgments.

Writing the Main Sections

Final Organization

Now that you have written an abstract and an introduction and have drawn up an outline, you are ready to write the main body of the manuscript. If your subject was already well known and understood, your planning was mainly a scheme for assembling the material into readable form. But if you had anything at all new or novel to report, organizing your ideas was probably a struggle. Now that you have fought it out with yourself in the outline stage, you have won more than half the battle. Perhaps without fully realizing it you have marshalled your forces according to the best principles of technical exposition: you have organized for logical order, arranged the headings and subheadings for the best clarity, thought about the relative proportions of the sections, and selected those topics you wish to emphasize. But the *final* organization, as already pointed out, depends on how you refine the outline as you write.

Data Sources

To follow your plan, you draw on the various sources of written information you have accumulated since the start of your project. You may even want to review reprints of papers that you had before the start of the project work. In your own work there are the entries in your engineering notebook, progress reports you and your colleagues have written, and possibly some patent disclosures. You may have prepared phase reports and analyses of special problems. You also probably have test data and performance records.

Using these various sources as a base, you will be paraphrasing and rewriting sentences and paragraphs that fit into the scheme of your outline. Because that scheme is a logical sequence of ideas, your writing can now begin to flow. But you undoubtedly have far more material than you could sensibly include in the manuscript. Even with a sound outline, well reasoned and well structured, you are faced with one other important decision, i.e., determining the proper amount of detail. Too many details defeat their purpose because they obscure your important points in a mass of minutiae; too few details lead, of course, to another kind of obscurity—lack of technical information. The choice of a middle road can be a delicate decision and deserves some further discussion.

Level of Detail

The best way to choose the right level of detail in your various sections is to adopt a length according to the purpose of the manuscript and the needs of your readers. This is similar to the choice of details for the introduction (Chapter 7) except that for the main body of the paper you are applying these judgments on a much larger scale.

To illustrate how you decide the question of detail, we consider four aspects of writing engineering manuscripts and show how the choices are dictated by the type of publication.

The rationale for design. Any paper on engineering design should give at least a hint about the author's reasons for choosing the design method. The author might compare the design approach with alternative methods and show the uniqueness and the advantages of the new approach. The level of detail will depend on the readership.

In a thesis the design rationale is important evidence of scholarship and originality. Because length is seldom a limitation, the author may devote several pages to this subject. In a journal paper two or possibly three paragraphs would be appropriate. In a technical trade journal article, one or two sentences would probably be adequate.

The amount of work done. In certain kinds of engineering writing, it is advantageous to demonstrate the sheer quantity of work done on the project. Such a manuscript may contain many tables and graphs, minute descriptions of procedures and equipment, voluminous test data, detailed analyses of problems encountered, engineering specifications, and so on. One place where such details are welcome is the internal report that summarizes work on a project and is intended for archival purposes. The data may be useful for a future project and may prevent useless duplication of work. A thesis on a topic that requires this kind of supporting data may adopt a somewhat lower level of detail. But in a journal paper or an oral presentation at an engineering seminar, you must not attempt to get credit for every bit of work done in the laboratory. For journal editors and referees, superfluous detail is intolerable. Overly long papers are expensive to publish, difficult to follow, and boring to read. To choose a suitable level of detail, look through the papers in a recent issue of the intended journal and use them as a general guide.

The relation to external literature. For a research or development manuscript, the nature of the contribution is customarily explained in the framework of the existing literature. For a thesis or dissertation or any type of review paper, this relationship should be set forth in rigorous detail. In most journal papers, only the more pertinent references are needed. In oral papers and trade magazine articles, only a few references, if any, are appropriate.

Visuals. In any oral presentation of an engineering topic, photographs and graph charts are usually helpful, and a large number of slides, foils, or flipcharts are seldom objectionable. In journal papers and magazine articles, the requirements tend to be different. Unlike the oral paper, where slides are shown successively on a viewing screen, the figures in your journal paper are laid out on the printed page in some relation to the text. Consequently, the published paper requires more care in the selection of artwork for proper emphasis of the author's main contributions. In a journal paper a smaller number of visuals (but often with more detail) is usually appropriate. A rationale for the choice of illustrations is offered in Chapter 13.

Making a Strong Case

The strongest parts of your manuscript are the main sections. These are validated by the other sections: abstract, introduction, conclusions, and appendix. To make the most convincing case for your engineering contribution, you must maintain proper relative proportions among the sections in the main body.

When you are writing from an outline, it is not always easy to retain your original perspective of relative proportions. Some of the sections are easier to write than others and tend to become longer than you intended when you were drawing up the outline. Other sections, although far more important, may be difficult to write, e.g., because of the complexity of a new concept or because of the problem of selecting the best technical level for your particular readership.

You should devote adequate length to the more significant portions of your paper—avoid overdoing those sections that are easier to write. An illustrative example of how misplaced enthusiasm can stretch a manuscript out of shape is as follows.

Refer again to the outline for your solar energy paper in Chapter 4. Section III.C deals with the heat transfer fluid. You may have a good deal of data on your experiments to determine the ideal ratio of propylene glycol/water, but it turns out that the choice of fluid has minimal effect on system efficiency. You can easily be carried away in your zeal to explain the algorithm developed for Section III.C. Unless you exercise restraint in writing that section, it will eclipse the important Section III.A on the design of the collector. One obvious way to keep the spotlight on Section III.A is to relegate the algorithm to an appendix.

Another way to maintain the strength and balance of your manuscript as you are writing it is to anticipate the content of your conclusions section. In the conclusions for the above example, you would show why the collector array is an ingenious design of high utility and would point out that the improvements due to the new coolant are minor. Looking ahead toward your final comments in the paper will remind you to keep your various sections in true perspective.

Chapter 9

How to Write an Effective Concluding Section

Ways of Concluding

For several important reasons, writing the concluding section can be putting your best foot forward. One reason is that a well-written set of conclusions brings your main contributions into focus. Another is that your concluding section has a more "captive audience" than the other parts of your manuscript. Many readers will scan only portions of the main body, but will carefully read the last section (which may be a summary, a conclusions, or a recommendations). One more reason is that most readers will *expect* you to explain the net results of your work in brief, readable form. Don't disappoint them. A strong closing brings your paper or report into the best perspective.

There are several ways for you to conclude. Your choice is important and will depend upon the kind of manuscript. To end your manuscript with the wrong kind of a concluding section is a poor strategy.

One standard way of concluding a paper is to summarize the main points, but a summary is appropriate only for a descriptive or tutorial manuscript. It should give more information than does an abstract or an introduction. For research or development reports it is much better to write a conclusions section. This kind of closing not only summarizes but also discusses the significance of the results. If your topic is an investigation of an engineering problem or an analysis of an existing situation to find the best course of action, a recommendations section is the obvious ending.

My recommendations for a strategic choice of a concluding section are summarized in Table 3. The following sections show by example how to conclude in the three ways marked with an asterisk in Table 3. For illustrative purposes I have used once more the idea in Chapter 4 for a manuscript on a solar energy system. If this particular manuscript were written as *a description of engineering features* of the system and took the

Table 3 The types of concluding section best suited to various kinds of papers and reports

Type of manuscript	Summary	Conclusions	Recommendations
Technical progress report	x		
Summary project report		x	
Proposal for engineering project*			x
Description of engineering features of device or system*	x		
Original design of device or system*		x	
Development of process or method		x	
Study of existing technical problem			x
Tutorial on engineering principles	x		
Review of developments in an engineering discipline		x	
Analysis of faulty procedures			x

* An example of this type of concluding section is given in this chapter.

form of an internal report, it would end with a summary. If it were written as *a design paper* for a journal, the ending would be an evaluating conclusions. And if it were written as *a project proposal* for building such a system, the report would end with recommendations.

The Summary Section

You may decide that your manuscript should end with a summary instead of a conclusions or a recommendations section. A summary must never be a weak attempt to review the main points in order to arrive at a graceful conclusion, i.e., to let the reader know that the end has come. Indeed, a perfunctory "bedside manner" at the expiration of a manuscript is no triumph for the author. A good engineering manuscript should instead end on a lively note, attuned to your main points in a vigorous and convincing finale.

Your summary must be discriminating and clarifying; by being selective it omits unwanted detail and gives your results in accurate perspective. The content, however, should differ in several ways from that of your abstract and introduction. Chapter 6, for example, suggested that an ideal abstract contains at least the following:

- a statement of the problem

- an indication of the approach to that problem
- the principal result

In Chapter 7, the ideal introduction included:

- the purpose
- a definition of the project goal
- the background
- reference to related papers
- the scope of the manuscript, etc.

By contrast, your summary should concentrate heavily on your results, such as the important features of the device or system, the accomplishments that led to the completion of your project, a recapitulation of the engineering principles, etc. This section should be brief but potent.

When your summary contains nothing of importance, it gives the reader a negative impression: The author has exhausted his literary resources and doesn't know what else to say; he is repeating the obvious; he has run out of ideas and somehow is trying to bring his paper to an end as best he can. Here is a brief example of such a summary:

SUMMARY

The SOLRAY system described in this paper is the result of an extensive design program carried out at the Solar Laboratory during the past three years. The combined efforts of the three teams of electrical, mechanical, and chemical engineers have produced a prototype system that shows promise for the future of solar heating. The field performance of the SOLRAY system meets the original engineering objectives of 13.5% thermal efficiency.

This example lacks focus, offers no important technical information, and does not summarize the results.

Some authors enumerate their results in a list. The items can be arranged in order of importance, for example, in a report on the features of a mechanical device. For other manuscripts, running text may be preferable to a list. Textual material can be more interesting because the sentence structure permits emphasis of certain ideas and subordination of others.

For an example of a good, informative summary you might consider once more a hypothetical report about a solar energy system that describes only its operational features and performance. You could summarize it as follows, without analysis, commentary, or recommendations:

SUMMARY (for a descriptive report)

The SOLRAY solar energy system has two unique design features that have contributed to the improvement in thermal efficiency. The main contribution is due to the novel configuration of the collector array. A secondary contribution is due to the adoption of a modified transport fluid for the transfer of heat from collector to storage tank.

Unlike conventional light-collector arrays, the new design concentrates radiant energy in two stages. The first utilizes a hyperbolic mirror to collect sunlight, and the second, a bank of Fresnel lenses placed between mirror and the absorption coils. A 10% rhodium-90% silver coating gave optimum reflective properties for the mirror. A bank of Fresnel lenses 1.2 cm in diameter, spaced at 1-cm intervals in a three-tier, linear array, provided the most practical arrangement for second-stage energy concentration.

The selection of 64% propylene glycol/36% water for the transport fluid was based on the physical and chemical properties of the mixture. The specific heat is 0.966 cal/g per °C at 60°C, and the corrosion rate of the galvanized iron piping system with the fluid at 97°C was only 6.03×10^{-7} g of FeO_3/cm^2 per mo. Field tests of the completed system yielded 49% efficiency for the collector area of 721 m^2 for a 35° angle at southern exposure. Overall thermal efficiency of 13.5% was obtained for the prototype system, as measured during the dynamic performance tests in Phoenix, Arizona, during July 1980. This result compares with the 9.7% efficiency previously obtained for the SOLTHERM system under the same conditions.

The Conclusions Section

When you are describing engineering developments and reporting novel designs or techniques, a summary is not always the appropriate way to end the manuscript. A conclusion, unlike a summary, not only reviews the results, but also interprets them. The most effective kind of conclusions serves as a candid critique of your own work, in which you point out what is important, what is significant, and why the results are valid. To make your manuscript most convincing and credible, you may even include some negative aspects, showing the gaps in your work and the limitations of your findings or your design. Many development papers also point out directions for future work and applications.

The following example might serve as a suitable closing section in a journal paper on the design of a solar energy system.

CONCLUSIONS (for a design paper for a journal)

The design feature that contributed most to the high overall efficiency of the SOLRAY system is the dual-concentrator collector array, which utilizes the principle of multistage concentration of light rays. This novel design differs from the two conventional configurations, i.e., (1) the flat-bed tracking array using a parabolic mirror and (2) the collector array using Fresnel lenses. The new design reported here provides a 49% energy gain by concentrating radiant energy in two successive stages, first with hyperbolic mirrors and subsequently with an array of Fresnel lenses. The analysis in Section II shows the theoretical limits of energy gain with mirrors and lenses. The design described in Section III provides a gain amounting to nearly 94% of the upper limit.

The second contribution to overall efficiency was the choice of the best proportion for the mixture of propylene glycol and water for the transport fluid. We used a modified Strauss algorithm, which included parameters for heat capacity, operating temperature, and corrosion of the collector coil material. This algorithm was superior to the original Strauss method, which had neglected the corrosion factor. The ideal proportion for the present system was found to be a 64/36 solution of propylene glycol/distilled water. During the dynamic tests of the system for 850 hours, no measurable corrosion or leakage occurred in the collector coils and pipes.

The prototype model of the system described here showed an overall thermal efficiency of 13.5% in the July 1980 tests in Phoenix, Arizona.

Analysis of the initial test results indicates that the final design modifications for the production model will require two changes in materials specifications. The first is the alloy used for the superstructure of the supports for the Fresnel lens array. These supporting members, which are 14 Fe–86 Mo rods welded to the bed of the collector array, conduct an appreciable quantity of heat energy away from the coils. It is anticipated that a superstructure made from an alloy having a lower thermal conductivity will raise the overall system efficiency by at least another 0.2%. The second change in materials specifications is that of the Cu–Al alloy used for the collector coils. The high operating temperature of 97°C, due to the dual concentration of solar radiation, brings the Cu–Al alloy just above the beta-phase transformation, drastically lowering the Young's modulus. A modified alloy composition having a higher transformation temperature will be selected for the production model to ensure long-term mechanical stability of the collector array.

In this example, the author, in a way, acted as his own technical referee— appraising the quality of the work, pointing out some minor defects, and suggesting how they can be rectified. This kind of concluding section strengthens the manuscript and makes it more believable.

At the end of a concluding section, it is customary to give credit for the support work that is normally a part of any large engineering program. This may consist of service work, such as analysis of materials, computation of design parameters, or testing of pilot models. Or there may be software support or applications development. In an internal report, this kind of engineering effort can be reported extensively and numerous acknowledgments can be given. In a paper given at a technical conference or published in an engineering journal, mention of support work should be minimized.

The Recommendations Section

Writing a recommendations report is an opportunity to demonstrate your analytic skills, the soundness of your reasoning, and your expertise in deciding on a course of action. The character of the final section will depend on the purpose of your report, e.g.:

- To analyze a problem, derive a solution from assembled facts, and propose a remedy
- To select a new procedure or technique and indicate why it is preferable
- To identify a need and suggest a way to fulfill it
- To propose a new project and show why and how it should be carried out
- To explore a new concept and recommend how it should be applied to existing engineering problems

Any reports of this kind should include a concise, clear-cut set of recommendations, the style differing somewhat from the usual conservative tone of engineering writing, which tends to include qualifying statements. When you decide on a course of action, do not be overly cautious. After you have established a rationale for your choices, make a firm, persuasive recommendation.

Suppose, for example, you were writing a proposal on how to modify a solar energy system to improve its thermal efficiency and lower its cost. Your argument for modification needs to be convincing. For example, you might prepare the final section in the following way.

RECOMMENDATIONS (for a project proposal)

This report indicates how recent advances in the technology of collector array optics and in the techniques of heat transfer can be used in the redesign of the SOLTHERM solar energy system. The radical nature of these design modifications and the preliminary tests of static thermal performance suggest a substantial improvement in overall thermal efficiency.

The estimates of dynamic operating characteristics given in Section IV predict an overall efficiency of 13.5% compared to the established figure of 9.7% for the SOLTHERM system. The proposed unit construction of the new collector array would effect a saving of 29% over conventional arrays in materials and assembly cost, as itemized in Table 2, Section III.

My recommendations for design modifications in the new system are summarized as follows:

1. Utilize the new two-stage technique of light concentration in the design of the collector array. This method employs hyperbolic mirrors and banks of Fresnel lenses for a 49% improvement over the performance of the SOLTHERM concentrating collector.
2. Adopt the modular assembly of 14 Plasticon Fresnel lenses fabricated by injection molding to replace the individual lens mounts used in the SOLTHERM system. The one-piece construction and the novel design of this bank of lenses are the basis of the cost advantage and the enhanced thermal efficiency.

3. Adopt the 10% Rh–90% Ag coating for mirror backing. This alloy provides three percent more reflectivity than the conventional silver backing.
4. Use the corrosion-resistant alloy of 75% Cu–23% Al–2% Zn for the collector coils. Because the beta-phase transformation of this alloy is well above the operating temperature of 97°C, the alloy meets the specifications for mechanical stability as well as corrosion resistance at that temperature.
5. Adopt a 64/36 solution of propylene glycol/distilled water as the standard composition of the transport fluid. The analysis in Section IV shows that this composition has the maximum specific heat and the best physical properties at the system operating temperatures.

The proposed design features of the new system are especially significant because the 9.7% efficiency of the SOLTHERM model had offered only a slight cost advantage over conventional furnace heaters for dwellings in the sunbelt region. The projected improvements in cost and efficiency of the solar energy plant described in this report indicate an important commercial potential for residential heating.

One of these types of concluding section is usually placed at the end of a manuscript, particularly if it is to be submitted to a journal. In industrial reports the summary or recommendation is sometimes placed instead at the beginning, and the abstract may then be omitted. The choice of placing this section at beginning or end may be either a stylistic preference or a requirement of your management. In any case, the distinctions among the three types of sections are important, as shown by the three examples.

Writing any concluding section is a demanding test of your ability to decide what to emphasize and what to deemphasize, which is the general subject of the next chapter.

Chapter 10

How to Achieve
Proper Emphasis in Writing

The Question of Relative Importance

When properly used, emphasis gives power and direction to your engineering manuscript. When misused, it not only weakens your writing but also disrupts the information flow that is so necessary in any technical report or paper.

You would not intentionally emphasize the wrong portions of your manuscript. As pointed out in Chapter 8, you might be "carried away" by your zeal to explain one aspect of your work that is interesting to you but is not your main contribution. In your misplaced enthusiasm you may devote a lengthy section to that aspect—say, an algorithm you developed—and thus attract undue attention to it. Such a long section would serve only to sidetrack your readers from your main theme instead of supporting it.

For effective engineering writing, you must give prominence to the portions you consider most important. The question of relative importance depends on your purpose. If your purpose is to *inform* (as in a report or journal paper), the strongest technical points in your manuscript are the items to be emphasized. If your objective is *to influence* your audience (as in a recommendation report or proposal), those items that will fill a specific need for your readers are the ones to emphasize.

Specifically, you could select one of the topics listed on p. 58 for emphasis, according to the type of manuscript.

It is clear, then, that while you are planning and writing your manuscript and anticipating that certain portions of it should be emphasized, you need to decide: (1) on the topics that are relatively important (i.e., to display the strong points) and (2) on the purpose of emphasizing them (i.e., to inform the reader of your strengths or to recommend a course of

57

action). You can then consider the various methods for achieving emphasis in your manuscript.

Manuscript	*Topic to be emphasized*
R & D paper	The main point of originality or novelty
Trade magazine article	The design features and utility
System description	The most important functions
Review paper	The significant achievements reported
Progress report	The goal-oriented advances made during the reported interval
Analytic report	The chief findings
Proposal or recommendation report	The facts and rationale that will motivate the reader

The Techniques of Emphasis

There are many ways for you to indicate that certain facts and ideas are more important than others. You can be forceful and obvious by loading your paper with illustrations, tables, and lengthy, enthusiastic explanations that leave little to the reader's imagination as to what you consider important. Or you can be subtle by trimming, subordinating, or even deleting the peripheral information so that your main ideas can assume a commanding position in the manuscript.

There are no rules for choosing among these techniques of emphasis. Your choices will depend upon your personal style of writing and the purpose of your manuscript. For example, if you are writing a research or development paper, you may rely mainly on structural methods to achieve emphasis—relative length, order, tabulation, etc. For a descriptive article on a device or system, you may prefer to rely on visual techniques, such as diagrams, layout, and color. For recommending a course of action, you may use rhetorical devices for emphasis, such as subordination or comparison. Or you may use any of the three methods for strengthening your manuscript. Whatever methods you choose, be sure to identify the chief contribution in plain and unmistakable language.

Structural Methods

Relative length. If you are reporting the engineering development of something new—perhaps a device, system, process, material, or measurement method—the most significant section of your report becomes more emphatic if it is either shorter or longer than the other sections. "Relative length" refers here to the amount of information you decide to include in the section and not merely the number of words and sentences. You can force the attention of readers by writing a short, pithy section about your main points. Or you can emphasize the technical aspects of your contribution by expanding that particular section, but only by adding *pertinent details* and not by indulging in verbiage.

Omission. Achieving proper balance in your manuscript is more important than packing it to the brim with technical information. Omitting some of the lesser details (or even some of the sections) can be enlightening, like the burnoff of morning fog, for clarifying the atmosphere. "Defogging" by eliminating detail is applicable to any unit of English composition— any phrase, clause, sentence, paragraph, or section of your manuscript. An engineer usually knows what to include in his or her writing, but knowing what to omit is also a prime requisite for producing a strong, effective paper.

Order. One interesting way to achieve emphasis is to decide on the best order for presenting your ideas. Tichy[59] suggests that in a short report the topics may be arranged in order of increasing importance to sustain interest; in a long report an arrangement of decreasing importance ensures that readers have the main points even if they do not read the whole report.

Position. An obvious way to draw attention to any given topic is to place it in a prominent position in the text, e.g., at the beginning or the end of a paragraph or section.

Tabulation. Tables of data can be used either to emphasize technical information or to deemphasize it. The usual purpose is to show trends and to demonstrate relationships, in easily readable compact form. But an alternative purpose of tables is for deemphasis, e.g., to provide a repository for compiled data, perhaps in an appendix, for convenient reference. The criteria for effective tables are reviewed at length in Chapter 14.

Lists. Another form of tabulation is the list, which merely enumerates items instead of comparing them. When set off from the text by indentation, the list forces attention.

Visual Methods

Headings. The headings and subheadings in your topical outline tend to be the key ideas in your manuscript. When these are used as titles of the sections in your finished paper or report, they alert the reader to the points of emphasis. Poorly chosen headings mislead your readers. In your manuscript draft you should review the section titles with great care for appropriateness and for accurate wording.

Typography. In certain kinds of engineering documents, particularly internal reports, you may have the opportunity to select type styles and font sizes. You can draw attention to illustrative examples by having them set in a smaller or larger type size than the rest of the text. Italics (or underscoring in a typescript) can be used occasionally for emphasizing a word or two. Such variations of typography retain their force only if used sparingly.

Layout. A different kind of opportunity for emphasis lies in the layout of a report. If, in addition to writing a report, you can contribute to its design, you can highlight the main ideas by placing figures and tables strategically. Another way to achieve emphasis is to avoid a crowded appearance in the page layouts. The proper use of white space helps to separate the components of your report and assign a prominent position to the material to be emphasized. When you are explaining a new concept or making a recommendation, the influence of eye appeal of the typed or printed page should not be overlooked. A topic can sometimes be given special treatment, for example, by placing it in a ruled "box."

Color. In graphs and charts color may be used for visual separation of the data and for emphasizing selected categories of information, but beware of the indiscriminate use of color. Strong, bright colors like red and orange might be used only for portions of your artwork that deserve special attention, with subdued colors used for items of lesser importance. If you are writing for a journal, do not plan to use color unless the publisher can accommodate it.

Figures. Charts, diagrams, and photographs not only attract the eye but also have a strong influence on the character of your paper or report. The novice tends to use figures liberally in an attempt to "dress up" the manuscript and to show the quantity of work done. The experienced writer gradually learns to use figures to their best advantage, i.e., to illustrate only the strong points in the paper. Criteria for figures are given in Chapter 13.

Rhetorical Methods

Subordination. Highlighting your significant ideas by deemphasizing the less important ones can be effective at every level of writing. Here is an example of a sentence with a subordinate clause:

> Because of the lubricating qualities of the colloidal graphite in the new formulation, Lube-all is ideal for packing wheel bearings.

The dependent clause about the graphite in the formula is subordinate to the primary fact that Lube-all is suitable for bearings.

This rudimentary form of emphasis also applies to paragraphs. Instead of relying only on a subordinating conjunction, you may begin with a clause that refers to the previous paragraph, as in the following example:

> Although these conventional lubricants have been adequate in the past for wheel bearings, an improved formulation like Lube-all is necessary for the new bearings in this year's automobiles.

The entire clause beginning with "Although" serves to subordinate the ideas about "these lubricants" expressed in a prior paragraph.

In the same way a short paragraph at the start of your most important section can be used to indicate that information in a previous section is subordinate.

Repetition. In the overall organization of your manuscript, you can emphasize the main engineering features or the chief results of your work by mentioning them in several strategic places. The primary contribution in your paper can be rephrased and reinterpreted in three different places (i.e., abstract, introduction, and conclusion) as illustrated in Chapters 6, 7, and 9. The examples given show how your main message can be repeated without using the same words and phrases.

Comparison and contrast. Another standard rhetorical device for achieving emphasis is to show similarities or differences. This is a variation of the technique I have labeled "subordination" and has the same purpose but a different form. Instead of using subordinating conjunctions, phrases, or sentences, it employs comparative or restrictive expressions to indicate relative importance. In the following example, the author's work is highlighted by comparing it and also by contrasting it with previous developments:

> The lubricant formula developed by Arby[13] has a base of highly refined Pennsylvania petroleum and an additive of two percent molybdenum disulfide, which is effective for bearing temperatures up to 178°C. The present formulation substitutes four percent colloidal graphite for the molybdenum disulfide constituent and retains the lubricating qualities for extended periods at operating temperatures up to 189°C.

Emphasis is given to the new formulation by citing its similar petroleum base and its different additive.

Word choice. The choice of appropriate wording is inherent in any kind of effective writing. The wrong word, when intended for emphasis, may not only be inappropriate but also misleading, for example:

> The enormity of the petroleum deposit was impressive.

The proper word having the intended emphasis here is not *enormity,* which means "monstrous" or "outrageous," but rather *magnitude,* meaning "extent."

Poor word choice, of course, is only one source of ambiguity in engineering writing. Human ingenuity in developing various shades of meaning for a given word is the source of much confusion, as is discussed in the next chapter on some practical implications of the science of semantics.

Chapter 11

How to Avoid Information Traps

A Pleasant Self-Deception

Many an engineer, satisfied with the first draft of a technical manuscript, indulges in a pleasant self-deception that the writing is clear and precise. Those who are most comfortable with the first draft are the least aware of the ambiguities that steal unnoticed into any initial expression of thoughts on paper or on the display screen of a word processor.

Pride of accomplishment is, of course, a great driving force when you are writing a manuscript; pride in your finished draft, however, tends to shield you from its defects in wording. Euphoria over your writing leaves you elated but uncritical. Your search for clarity of meaning demands disciplined thinking, unhampered by any strong emotion. In general, however, you can gain some insight into the treachery of words and their nuances by considering the mechanisms of multiple meaning and observing some examples of typical information traps.

Communication between author and reader can be disrupted by misuse of semantics and syntax. Because these two academic terms are of great practical interest to all who write, I offer simplified, pragmatic definitions:

- *Semantics:* The indirect relation between words and what they refer to.
- *Syntax:* The arrangement of words in a sentence.

As an engineering author you create technical documents to offer specific information to readers who need it; for this reason you must be sensitive to the proper use of semantics and syntax. Otherwise, you will unintentionally distort both meaning and significance.

63

The Meaning of Meaning

Good writers sense the problems of ambiguity. For some, precise meaning becomes an obsession. T. S. Eliot in *Four Quartets* described the agony of the writer as "an intolerable wrestle with words and meanings." Yet, as suggested by Lois DeBakey,[21] some of us are insensitive to the abuse of language and underrate the ability of words "to shape man's thinking, polarize his attitudes, mold his character, and dictate his actions." A poor choice of words weakens the power of a manuscript and obscures meaning.

Ambiguity and vagueness, however, can also be deliberate. A complete engineering project does not always provide answers to all the technical problems that were encountered nor a full understanding of all the results. The engineer writing a project report may not attempt to clarify uncertain details. But he or she should clearly distinguish between what is known and what is not known.

Most words, of course, have multivalued meaning. Moreover, as shown by Ogden and Richards,[49] an important concept in semantics is that the precise definition of a word depends not only on its context, but also on the personal background of the reader. Words are like chameleons, reflecting the color of their environment. But some words have only generalized meaning, which may be interpreted in any of several ways regardless of context, and others have definitions that shift with time.

Words are not necessarily related directly to their definitions. Figure 3 suggests an indirect relation between word, thought, and a physical object or idea. The connection between word and object is created in the mind of the reader who sees the word and associates it with some object or idea in his memory.

An example of the relationship in Figure 3 between symbol and referent would be the word "book," located at the left of the triangle, and a book (physical object) at the right of the triangle. The only direct connection between the referent and its symbol would exist in the associative mental process (the reference) at the top of the triangle.

A rare exception is the onomatopoeic word, such as "hiss." In this case the sound of the word is almost identical with the thing it represents.

The psychological interpretation of the association between word and thought is suggested in Figure 4. Here the stimulus/response reaction depends on the experience of the reader. At the sight (or sound) of a word, the stimulus S, there is an initial response R_0, from which the reader (or listener) extracts a specific association $R_1 \ldots R_n$. The writer's intended R_1 response may be "trapped" when the reader experiences an R_2 response instead, either because of uncertain context or because of personal reaction.

The word "slip," for example, has several engineering definitions. The referent depends not only on the context but also on the terminologies

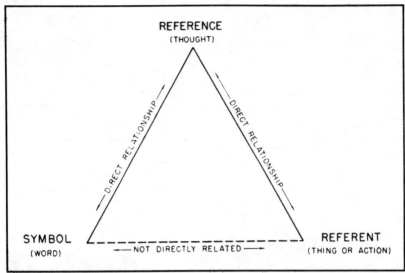

Figure 3   Relations between word, thought, and thing. (Reprinted from reference 39.)

familiar to a given reader. Various engineering specialists will have separate responses R:

To the electrical engineer	R_1 is the lag of an induction rotor speed behind the speed of the rotating stator field
To the metallurgist	R_2 is the displacement of portions of a crystal along parallel planes
To the ceramic engineer	R_3 is a mixture of clay and water
To the marine engineer	R_4 is a sloping ramp for repairing ships

These definitions are precise meanings of "slip" and depend on the context. The problem of *ambiguity,* due to multiple meanings, is quite different from the problem of *vagueness,* which is due to lack of specific meaning. Vague words are usually the overworked ones like *many, very, actually,* etc., and abstractions loosely used, like *management, reliable, system, optimum.* The usefulness and intelligibility of an engineering document can be inversely proportional to the number of vague words it contains.

To the layman, engineering and scientific languages may seem well defined and exact for the intended specialist. However, to the reader of technical literature the language is often far from unambiguous.

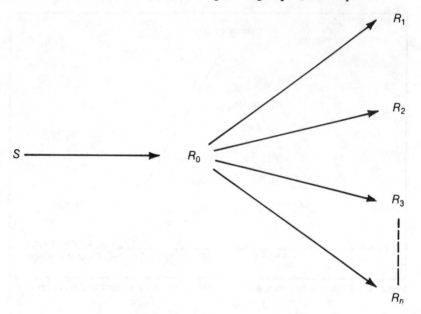

Figure 4   Stimulus–response interpretation of multiple meaning. (Reprinted from reference 42.)

Traps for the Reader

A few examples illustrate various kinds of semantic traps. The first is the subtle question of multiple meaning, where both writer and reader might be wholly unaware of the ambiguity:

> The next sintered composition for the experimental heating rod consisted of a mixture of talc with 23% zinc. This element proved to be unsatisfactory.

At first glance you may consider the meaning of these two sentences to be unmistakable, until you think about the possible referents of "element." Which was unsatisfactory, the experimental heating element or the chemical element zinc? A better construction would be

> The next sintered mixture, consisting of 77% talc and 23% zinc, could not be used as a heating element because its resistivity was too high for the ratio of talc/zinc.

The problem in the next example, concerning the use of abstractions, should be more obvious to you at first reading:

> We spent considerable time this month on the repair of the automatic stirring device before we could continue the study of solubilities.

In this example there is no adequate referent for "considerable." A supervisor or contractor, in evaluating the expenditure of project resources, would want to know how much time was spent on the repairs.

When terminologies are misused, the confusion of meaning may or may not be noticed, as in the following sentence:

> The density of the porcelain rod, after firing in hydrogen at 1100°C for three hours, was 2.32 g/cm^3.

The exact meaning of the word "density" (weight per unit volume) of a ceramic body may be either the true density or the apparent density, the latter being the measured value that is uncorrected for pore volume.

Syntax

Numerous examples of faulty syntax are cited in every grammar book. The distorted sentences are frequently incongruous, or even comical, but not necessarily ambiguous. In engineering literature the more troublesome cases are those twisted constructions that are downright confusing or, worse still, the ones that are subtly misleading. The examples given here illustrate the potential information traps caused by poor construction of a phrase, clause, or sentence.

One problem in syntax is the squinting modifier, placed between two parts of a sentence so that the reader cannot tell which part is being modified:

> After it was decided to use the new buffing compound, because of the weaker acid content, the surfaces attained a mirror finish.

The writer's implication is not at all clear. Was the compound chosen because it happened to have a weaker acid content, or was a mirror finish obtained because of the lower acidity? The following is an improved version:

> The weaker acid content of the newly chosen buffing compound gave the surface a mirror finish.

Dangling infinitives, misplaced phrases, and obscure referents can cause a combination of troubles like the following:

The shop foreman was asked to start the shift at nine o'clock.

The reader is confronted with simultaneous information traps: Was the foreman asked at nine o'clock? Was he asked to start with his own work at nine or to start the entire group of workers on the nine o'clock shift?

One solution to the entanglement of misplaced words, modifiers, and referents is to divide the sentence into two or more separate statements. But sometimes it is necessary instead to clarify two confused sentences by combining them. The dual meaning in the following sentences, for example, could entrap the unwary reader:

> The full shipment of these machines has not been received. The last delivery date was May 4.

Some readers will assume that the last time some of the machines were delivered was May 4. Others would understand that the most recent promise for delivery of the remaining machines was made on May 4. The writer might attempt to revise the delivery date statement in the following way:

> Not all of the machines ordered have been shipped. The supplier promised to deliver the remaining machines on May 4.

The new version may seem improved until you consider the significance of May 4. Is this the promised delivery date or the date the supplier made the promise? One solution is to combine the sentences:

> May 4 is the most recent delivery date promised for the remainder of the shipment of machines.

These examples suggest that ambiguity and vagueness are not always immediately obvious to the reader, who must be on guard against the kind of information traps shown. And the author must find every way to protect the integrity of the manuscript from the start.

Traps for the Author

As an engineering author you need to develop specific methods for avoiding the pitfalls of distorted information. Shakespeare seemed to be poking fun at the engineer caught in his own trap when he wrote, in the third act of *Hamlet,* "For 'tis the sport to have the engineer hoist with his own petar."

Try these four ways of minimizing the traps as you write and of eliminating those that have slipped in later because of your uncertainties in semantics and syntax:

1. *Define your terms.* You can do this either by inserting definitions of specialized words the first time they appear or by appending a glossary of selected words and phrases.
2. *Wait a while before reading your draft.* Put your finished manuscript aside for a day or two before attempting to read it critically. A fresh, new look at it will turn up some incongruities and distortions you had not noticed while writing.
3. *Have your colleagues review the manuscript.* The separate readings from various points of view will help you find the trouble spots.
4. *Spend a reasonable time on revision.* A hastily written and briefly revised manuscript may be loaded with ambiguous and vague expressions. Frequently, the difference between a poor manuscript and an excellent one depends on efforts to get independent reviews and on thoughtful revision.

One way to minimize information problems in significant areas of your manuscript is to support your text with visual aids in the form of graphics and tables, which are the subjects of Chapters 13 and 14.

Chapter 12

Writing Productively
with a Personal Computer

In the Driver's Seat

Practicing engineers and students are now finding many applications for the personal computer, which was originally developed and designed by engineers. The PC can be used not only for word processing but also for generating graphics, for statistical analyses and calculations, information retrieval, electronic mail, and so on. You can utilize all of its various functions for composing engineering papers.[3] However, your main interest for manuscript preparation will be word processing.

Because of the rapid proliferation of computers in engineering firms and in the schools, your decision on whether to use one may be mainly a matter of personal preferences. We are all creatures of habit. If your long-standing habit is writing by hand, you may feel more comfortable in generating that kind of manuscript. Or you may habitually knock out rough drafts on the typewriter, perhaps not as a typist but as a hunt-and-peck writer. Why change your habits?

Today's rapidly advancing technologies steer us toward new work habits.[4,22,66] Your first approach to using a personal computer may be frustrating for a while until you get the feel of it and learn the operating procedures. Within a short time, however, you will find that corrections to the text are easier and faster than when writing by hand or using a typewriter or even an electronic typewriter. The flexibilities of the PC encourage you to write, and also you are more likely to review and revise your work.[16,33,56]

But less obviously the PC influences your work in other ways. You may not realize that, because you spend less time making revisions, you maintain better continuity of thought. Indeed, the psychology of composing on the PC differs from that of drafting a manuscript by hand or on a typewriter. Because of your interaction with the computer you also develop

70

a new attitude toward expressing your ideas on paper, and your writing habits undergo some subtle changes. For example, fleeting ideas are a part of every writer's composing process, and the computer permits quick recording of the text before you lose it from short-term memory.[18] By reducing the time required to record your ideas, the machine permits you to spend a higher percentage of your time on refining the technical content of your manuscript. Also, writer's block tends to disappear because using the machine forces your attention and invites you to take control.

Seated at the PC keyboard you are in command, giving instructions for the computer to swing into action! Unlike the ball-point pen or the typewriter, the computer offers you a choice of a great variety of automatic operations. As a start you can specify the format (indentation, line space, margins, etc.) and also the type fonts, if you have a dot-matrix printer. As you compose you can rapidly erase letters, words, paragraphs, or whole pages. You may rearrange or insert words in a sentence, and even shift sentences or transpose paragraphs instantaneously. Certain word processors provide automatic tables of contents and indexes. You can use a split screen or "window" for comparing or merging separate parts of a manuscript. Many word processing programs now include spelling checkers, and some will even analyze your writing style for wordy phrases, trite adjectives, and grammatical errors.[14,34,35,45] Some programs provide the synonym check, illustrated in Figure 5. Others generate line drawings and graph charts. With a modem connection to an information service you can do literature searches.

The various commands made with a few keystrokes give you the satisfaction of rapidly controlling your writing output. *This puts you in the driver's seat!* You then have a new sense of "movement" in the writing process.

Interaction with the Machine

You communicate with the machine by giving it coded instructions. Many programs available for a personal computer will communicate with you by displaying messages on the screen, telling you what to do, offering you choices from a "menu" of computer actions, informing you that you have made a mistake, or advising you of the status of its operation.

For example, it may take a few seconds for the PC to read through its dictionary of many thousands of words to operate the spelling checker. Meanwhile the following message may appear on the display screen:

Word list being read—please wait

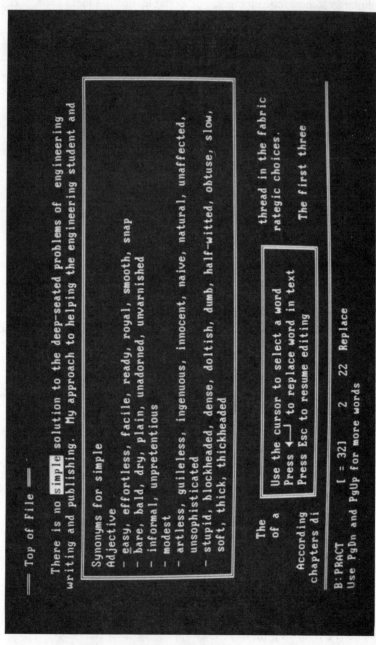

Figure 5 An example of the synonym check, provided by the Word Proof program. In the case illustrated, the program offers 35 synonyms as potential replacements for the highlighted word "simple."

This instructs you not to press any keys until the computer is ready. Or if the machine senses that you may have made a mistake in requesting the erasure of an entire file, the following precaution can appear on the screen:

Delete is permanent—Are you sure (Y/N)?

You then type "N" for "no" if you did not intend your whole composition to be erased from memory of the system unit.

And so interaction with the computer is a two-way street. Not only do you tell it what to do and how, but from time to time it also raises questions to prevent you from straying from the correct path.

In addition the system actually becomes an aid to your reasoning power as you explore and test your engineering ideas with computer programs. With these you can build models and do simulations to confirm and extend new technical concepts. This kind of analysis and synthesis is limited only by the programs available to you for this purpose and by your ability to adapt them to your use and absorb information from the machine's results.

Interaction with People

When your PC is connected into a communication network, it gives you the opportunity to react with other people. You may have one or more co-authors at remote locations, each preparing portions of the manuscript. If their personal computers are connected in the network, you may type comments and suggestions on your keyboard and send the messages to them by electronic mail. You can also exchange texts with them electronically.

This kind of quick interchange with your colleagues on the network is a decided advantage in any cooperative writing project. Trading notes on your manuscript by instantaneous transmission provides the inputs you need for rapidly evaluating the manuscript and improving the writing quality.

Incremental Writing

The method of writing in increments described in Chapter 5 is conveniently carried out on the personal computer whether it is operated stand-alone or connected to a host computer. From time to time, as your project work proceeds, you enter documents (including data from your engineering notebooks, reports, and memoranda) into the computer, where they are stored either on a diskette or in a database in the host computer.

When you start to prepare your engineering paper, you can then retrieve any of those documents, either for display on the screen or for insertion into your manuscript. The computer storage then becomes an extension of your memory. W. J. Doherty[23] suggests that the thought processes you enter as data in storage are a form of "captured intelligence," available for instant retrieval. Also, when you prepare a manuscript piecemeal in this way, you can alter it quickly according to ongoing changes in the engineering project.

You store reference data for instant access, with no need to search through existing papers and notes. Among the various items you can store and retrieve are bibliographic references, which you can select and quickly rearrange in any desired order with a conventional SORT program.

Your Decision

The advantages cited in this chapter are ample reasons for deciding to adopt the PC and a careful selection of software as your work tool. If you are to get the best use from it, however, you need to understand its limitations. It is essential to learn the basic operations of the machine and to find the right programs for your purpose. Spending a little time to learn the functions and the capabilities of your programs will pay off handsomely when you work on a manuscript. Special functions provided by those programs save you keystrokes, time, and energy.

After you have learned the limitations and the conveniences of word processing the manuscript, you will fully understand the real value of the PC and the stimulus to good writing that it provides.

Chapter 13

How to Choose Illustrations for Visual Impact

The Forces at Play in Graphics

The various forms of illustration—charts, diagrams, drawings, and photographs—can obviously have a strong influence on the character of your manuscript. A strategic choice of figures for your paper or report will greatly strengthen it. A poor choice will inevitably weaken the fabric of your writing. For these reasons you should not think of the figures as something separate from your engineering document, to be tacked on as an afterthought to support the text. Instead, you must realize that illustrations properly chosen before you begin to write can be a strong driving force in your manuscript rather than a crutch for weak exposition.

When writing for a general audience, some writers use figures merely to attract attention, to break up the monotony of solid pages of text, or to "dress up" the article. But when writing for peers who need engineering information, you should avoid any attempts to make your paper artificially palatable and to win an audience. For the interested reader those methods of forcing attention tend to dilute your message rather than enhance it. The most powerful way to attract readers, to break up the monotony of the text, and so on, is to provide the *impact* of illustrations that effectively bring out the content of your manuscript.

When designing the illustrations that will make the strongest contributions to your document, be certain to select the most effective form for each figure. Observe, for example, the distinctions among the four basic forms listed in Table 4.

Your choices are important. They can affect the readability of the paper and even its validity.

Table 4   Typical uses for the four basic forms of technical illustration

Form	Typical application
Charts*	
Line chart	Relationships between variables; comparisons; trends; relative changes
Surface chart	Cumulative totals of data from successive components
Bar chart	Comparison of fixed sums for successive time intervals
Pie chart	Relative proportions of the parts of a process or object
Flow chart	Steps in a process
Drawings	Pictorial representation of physical objects: details, cross sections, motion, functions, "exploded" assemblies
Diagrams	Abstractions; symbolic configurations; operating principles
Photographs	Realistic views of objects in meaningful perspective

* The general forms of charts are illustrated in Figure 6.

Weak Choices

The dominant role of illustrations becomes evident when you make poor decisions in their design and application. Your charts lose effectiveness when you choose the wrong form. A *line chart* (Fig. 6a), for example, does not serve its purpose if the coordinate scales are not sufficiently precise when you need to read data from the curve. Such data are better listed in a table if precision is important. *Surface charts* (Fig. 6b) are suitable only when the curves depict gradual changes, and such charts are ineffective if any of the curves overlap. *Bar charts* (Fig. 6c) do not show data clearly when the differences between the quantities shown are slight. Moreover, if the scales are expanded to show small differences, a bar chart looks suspiciously artificial. *Pie charts* (Fig. 6d) give a good picture of percentages of a whole process but are of little value for showing a large number of small percentages.

A *line drawing* can be used to illustrate details that would not appear in a photograph. But the real value of a line drawing is defeated unless it is designed for clarity and emphasis of the important pictorial features, such as cutaway or cross-sectional views, and to omit unneeded details.

Although a good *photograph* can be ideal, a poor one may lack desired details in the highlights and shadows. Or the picture may have a confusing background. Another limitation is reproducibility. Even an excellent photograph will not always reproduce well in print unless the publisher uses a fine halftone screen and high-quality paper stock.

The overall quality of your document depends not only on these subtleties in the choice of the best form for each illustration, but also on other factors, e.g., the number of figures, the choice of topics, the amount of detail in the technical artwork, and the relation of figures to text. Poor

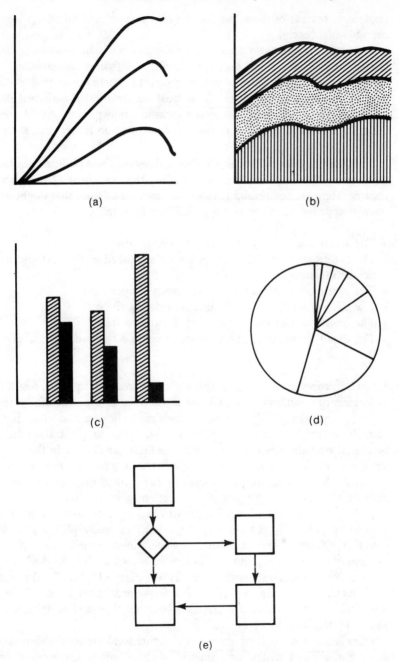

Figure 6 Some basic types of charts—(a) line chart, (b) surface chart, (c) bar chart, (d) pie chart, and (e) flow chart. Applications are listed in Table 4.

choices in several of these factors can combine to ruin an otherwise acceptable manuscript.

One of the more common problems, particularly in manuscripts submitted to journals, is an excessive number of figures. Illustrations are expensive to reproduce and editors prefer to delete those that do not contribute anything significant. You must decide what is sufficient illustration to give your readers adequate technical information. Avoid a profusion of figures that tend to diffuse your information rather than sharpen it.

When you delete unnecessary charts, diagrams, and photographs from your manuscript, you are probably ridding the document of some weak choices. Here are some valid reasons for omitting a figure that might have seemed appropriate when you prepared the first draft:

- The figure has already appeared in the literature.
- The information in the figure is amply explained in the text and need not be illustrated.
- The figure emphasizes a trivial or irrelevant point.
- The data properly belong in another existing figure.
- The data could be more effectively listed in a table.
- The details in the figure serve more to confuse the issue than to clarify it.

The proper grouping and spacing of figures is frequently neglected by engineering authors, especially by those who have not carefully organized the manuscript. My experience leads me to believe that when figures "bunch up" unpleasantly in the page layouts, so that an illustration cannot be placed near its mention in the text, the fault usually lies in the overall structure of the document. When you write a paper or report, try to visualize where the figures will appear on the printed page, and adjust the number of figures and lengths of text sections accordingly.

Properly constructed, your paper will highlight its strongest and most interesting points by illuminating them with carefully placed graphics. Good coordination with the text, however, involves not only effective placement of the figures, but also proper reference to them in the body of the manuscript. A glib statement, such as "see Figure 16," can be altogether inadequate. It is better to explain in a few words what Figure 16 is or does, such as "This nonlinear relation between time and temperature is shown by the kink in Figure 16."

Including too much information in a figure weakens your presentation. A simplified chart or diagram, illustrating the essential points and devoid of minutiae, is usually more effective than a complex figure. In simplicity there is strength.

Strong Choices

To build vigor and conviction into your illustrations, use the existing strengths in the manuscript. Of course, it is not always easy to convert engineering concepts into clear, definitive graphics, and you may have to wrestle with the problem of designing figures that will have the greatest visual impact. Like any good wrestler you first have to estimate the strength and weight of your opponent and then use thrusts and twists that make him fall of his own weight. In the same way you look for the "heavy" topics in your manuscript, e.g., the new design, novel methods, innovative processes, improved operating features, enhanced performance, or whatever you feel carries the most weight. These topics then become your strongest choices for illustrations, depending on the kind of manuscript:

Type of manuscript	Emphasis in the illustrations
Engineering development paper or thesis	The new developments; the nature of your contribution; its significance
Tutorial article	The important concepts; their practical application
Descriptive report	The main features, details of structure and/or operation

The best way to make those choices was suggested in Chapter 4. Do your planning in the outline stage. Decide on central points that should appear in the figures. Then build up the manuscript around the proposed figures. This scheme will ensure that your illustrations become an intrinsic part of the manuscript and not a set of adornments added after the writing to create an air of respectability.

Selecting the most important topics, however, is only one criterion for illustrations that will give the most vitality to your manuscript. After you have selected the "goodies," be sure to design the figures for your intended audience. The concepts in Chapter 2 on matching your objectives with the readership apply to graphics as well as to writing. The artwork in a journal paper for your peers, for example, should include more technical detail than the figures in a report for engineering executives.

If the final version of your figures is to be prepared by a draftsman or technical illustrator, it is important that you provide instructions on the purpose of your artwork, the type of publication, and the audience. If you make the drawings or take the photographs yourself, consult with your colleagues on the best treatment of the figures. For charts and diagrams a third alternative is to use the workstation terminal, which is gradually becoming available to engineers who prepare their own manuscripts. If you have a terminal with a computer graphics facility, you can construct your own charts.

Some examples of how engineer authors have chosen illustrations for maximum impact are given in the following section.

Illustrative Examples

Excellent examples of well-chosen figures can be found in the engineering literature. I have selected a few that show how the author decided to illustrate the main feature of the manuscript in one or more figures. In each case the artwork presents in visual form some technical information that cannot possibly be conveyed by words.

The importance of such examples is obvious to the author of any kind of technical material and even more so to an illustrator. It was Leonardo da Vinci who commented on the value of technical art five hundred years ago in his *Notebooks*[9]:

> And ye who wish to represent by words the form of man and all aspects of his membranification, get away from the idea. For the more minutely you describe, the more you will confuse the mind of the reader and the more you will prevent him from a knowledge of the thing described. And so it is necessary to draw and describe.

Our first example of the necessity "to draw and describe" is the work of H. S. Black, who published a definitive paper[11] in 1934 on his invention of the negative feedback amplifier. This work was a major innovation and later found wide application in the entire field of electronics and communications. Three of the illustrations in Black's paper are reproduced here in Figures 7–9. In these three figures are the essence of his idea of negative feedback. Figure 7 is a diagram of the basic concept, easily understood at a glance by any electrical engineer. The details of the diagram are defined in the legend, making the figure fully self-explanatory. This illustrates an important principle. Although a picture is supposed to be worth a thousand words, most figures in engineering papers need a descriptive caption to be independent of the text. The legend in Black's figure is somewhat longer than usual, but the length is justified by the originality and complexity of the concept, which was drastically simplified in the diagram.

The idea of negative feedback, which reduces signal strength, needed a good bit of explaining because of its use in circuits that amplify signals. The role of the figures is to show how the mechanism operates and to give concrete examples of the experimental results.

In Figure 8 the author shows an application of the feedback. The circuit diagram is intended only to indicate how the feedback path connects

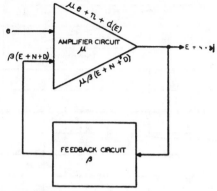

e. signal input voltage
μ. propagation of amplifier circuit
μe. signal output voltage without feed-back
n. noise output voltage without feed-back
d(E). distortion output voltage without feed-back
β. propagation of feed-back circuit
E. signal output voltage with feed-back
N. noise output voltage with feed-back
D. distortion output voltage with feed-back
The output voltage with feed-back is $E + N + D$ and is the sum of $\mu e + n + d(E)$, the value without feed-back plus $\mu\beta[E + N + D]$ due to feed-back.

$$E + N + D = \mu e + n + d(E) + \mu\beta[E + N + D]$$
$$[E + N + D](1 - \mu\beta) = \mu e + n + d(E)$$
$$E + N + D = \frac{\mu e}{1 - \mu\beta} + \frac{n}{1 - \mu\beta} + \frac{d(E)}{1 - \mu\beta}$$

If $|u\beta| \gg 1$, $E \doteq = \frac{e}{\beta}$. Under this condition the amplification is independent of μ but does depend upon β. Consequently the over-all characteristic will be controlled by the feed-back circuit which may include equalizers or other corrective networks.

Figure 7 Amplifier system with feedback. (Reprinted with permission from *Electrical Engineering* 53:114, 1934.)

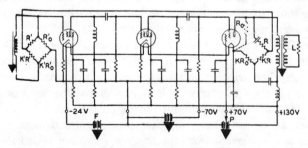

Figure 8 Circuit of a negative feedback amplifier. (Reprinted with permission from *Electrical Engineering* 53:114, 1934.)

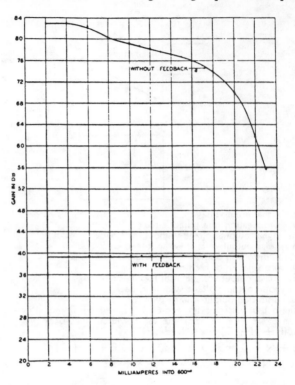

Figure 9   Gain-load characteristic with and without feedback for a low level amplifier designed to amplify frequencies from 3.5 to 50 kHz. (Reprinted with permission from *Electrical Engineering* 53:114, 1934.)

output with input. For that purpose all labels of circuit components are omitted except those that feed the signals back from the output. The figure gives only a general idea of an application and is not the most important contribution to the paper. Therefore, the figure is constructed in a small format, *even though its complexity would have justified making it much larger.* The more significant results are later emphasized in the paper by displaying them in charts of much larger size.

One such chart is reproduced here in Figure 9. This chart illustrates dramatically the exact nature of the improvement in the gain characteristics of an amplifier when negative feedback is used at various loads. The caption explains the variation of gain with load in a given frequency range and shows the effect of feedback without relying on the text.

The simplicity and clarity of these three figures demonstrate how illustrations can be derived from the text itself and not added merely to stimulate interest; how the greatly simplified charts and diagrams can still be highly significant and self-contained; and how such illustrations provide maximum impact by revealing the core of the author's message.

But you do not have to invent the wheel to generate effective illustrations. The same principle of concentrating on the main technical ideas applies not only to development papers, but also to descriptive articles and reports and to tutorial or review papers that discuss well-known engineering principles.

The second example is from a 1957 paper by T. Noyes and W. E. Dickinson[47] that described the first disk file, designed for computer applications. Publication of that paper revealed the principles of a device that stored information magnetically on rotating disks. Instead of using the conventional punch cards, magnetic drums, or magnetic tapes, the new system used disks mounted on a vertical shaft, vaguely resembling records stacked on a spindle in a phonograph. It would have been quite difficult to describe such a machine without technical illustrations.

The authors' choices are interesting. The figures portray details of a critically important device, called an "access mechanism," which is used to find any digital record at random on the disks. The authors used a photograph (Figure 10) to show the general appearance and a diagram (Figure 11) to indicate the mechanical configuration and the positioning arrangement. In the diagram, two encircled drawings effectively show small structures that would have been quite difficult to photograph. The advantage of the photograph in Figure 10, on the other hand, is its realistic view of the exterior, unembellished by an artist's interpretation. Thus the authors effectively used two illustrations (Figures 10 and 11) for the sole purpose of showing the operational features of the device.

In contrast, the author in the next example used a single illustration for two separate purposes. Figure 12, which shows the configuration of the memory cell and also the steps in fabrication, appeared in a paper by R. A. Larsen.[32]

The important contribution brought out in the diagram is the unique method of forming the microscopic oxide channel. First, in (a), boron ions are implanted on the gate surface, which is then oxidized and subsequently, in (b), coated with nitride. The figure thus simultaneously focuses on the method of precise fabrication and the features of cell design. Such a diagram is not merely an aid to understanding the text. Rather, it defines the complex forms that could not be explained in the text alone.

Another entirely different purpose of technical illustration is to show the result, in graphic form, of an author's new computational technique. The growing use of computer terminals now permits engineering authors to generate the kind of artwork shown in Figure 13, which appeared in a paper by Jones and Paraszczak.[29]

Figures 13b and 13c were prepared by computer graphics at the

Figure 10   Access mechanism of the IBM Type 350 magnetic-disk, random-access memory. (Reprinted with permission from *IBM Journal of Research and Development* 1:72, 1957.)

authors' desk terminal. The paper describes the use of a computer program to simulate, in three dimensions, the formation of very small holes in a polymer photoresist film that had been exposed to an electron beam and then chemically developed. The depth and shape of the holes etched in the film depend on the development time, which was calculated using the authors' algorithm and is shown in the figure. The computer drawings produced by the algorithm give fine details. The simulated profiles, for example, have the curvature at the corners characteristic of holes in photoresist bombarded by electrons and subsequently developed.

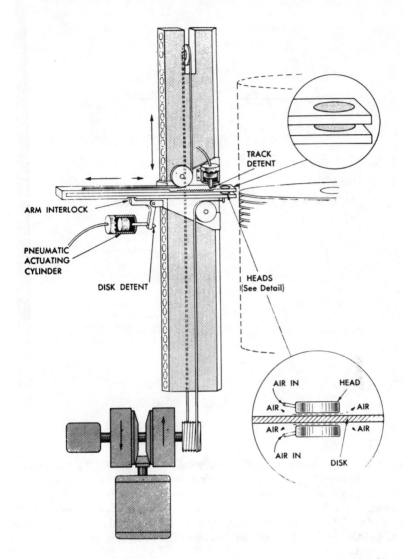

Figure 11 Functional diagram of access mechanism showing general scheme of vertical and horizontal positioning, interlock, and actuator group. (Reprinted with permission from *IBM Journal of Research and Development* 1:72, 1957.)

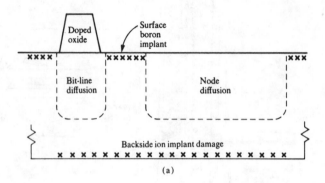

(a)

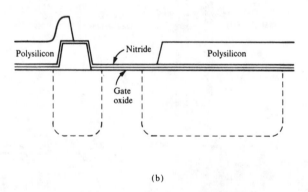

(b)

Figure 12   Views of the SAMOS memory cell. Figure (a) shows a cross section immediately prior to gate oxidation, while (b) is the same cross section after the polysilicon patterning mask step. The structure after the first-level metallization is shown in (c). A top view of two nodes on a bit line is shown in (d), with the extent of n⁺ diffused areas being shown by dashed lines. (Reprinted with permission from *IBM Journal of Research and Development* 24:268, 1980.)

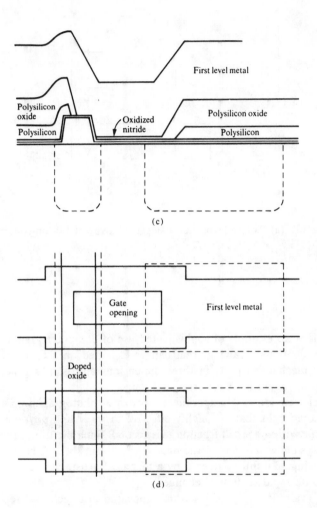

(c)

(d)

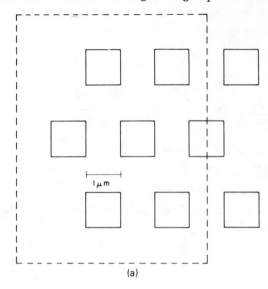

(a)

Figure 13   (a) Pattern layout for a staggered array of 1.0-μm contact holes.

This is an example of sophisticated use of computer graphics. Other more conventional drawings that can be executed on a terminal include complex mechanical parts in three dimensions and graphs generated directly from the author's experimental data.

Some engineering experiments provide photographs made with stroboscopic light that are highly effective in recording physical changes that occur within a small fraction of a second. Such pictures, however, do not always show the inner structures in the objects being photograped. A line drawing, skillfully rendered by a technical artist, can be used instead to illustrate rapid structural changes.

An example is Figure 14, which appeared in a technical report. The six parts of the figure depict a time sequence of events illustrating motion, force, and impact resistance. The "shaped round" is a projectile fired at a laminated armor plate. The series of drawings shows the approaching projectile, the force fields, the disruption of material in the plate, and its resistance to penetration.

A line drawing can thus show physical action and its associated concepts (such as shock waves) that cannot be fully described by either words or photographs.

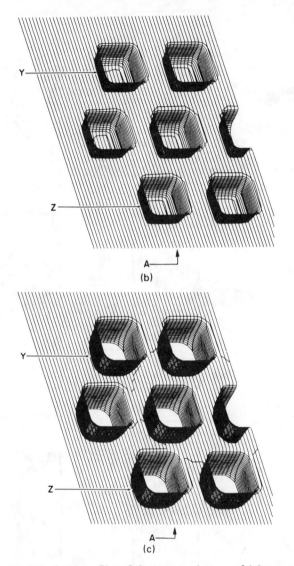

Figure 13 (b) Calculated profiles of the staggered array of 1.0-μm contact holes after a development of 70 s.   (c) Calculated profiles of the staggered array of 1.0-μm contact holes after a development of 110 s. (Reprinted with permission from *IEEE Transactions on Electron Devices* ED-28:1544, 1981.)

These five examples show how an author can use figures forcefully to represent the main ideas in a paper or report. Such skillful use of technical illustration can reveal and reinforce the strengths in an engineering manuscript.

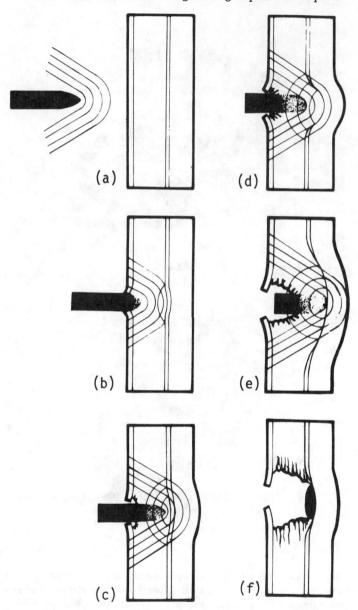

Figure 14   The impact on a laminated armor material by a shaped round. (a) Shock waves associated with the projectile; (b) shock waves entering the laminate; (c) start of shock reversal and energy dissipation; (d) energy dissipation at midpoint; (e) energy dissipation at advanced stage, with spall shield action; (f) defeat of projectile. (Reproduced by permission of Vivian Drury, ART 'n' Type, Scarsdale, N.Y.)

Chapter 14

How to Construct Tables
for Information Content

Technical Content

Before you attempt to draw up a table of data for a paper or report, consider the purpose of the tabulation. Formal tables are used for two different reasons. One is to highlight a topic that is new or significant. The other is to provide a compilation of reference data. How you design the table and where you place it thus depend on what you expect it to do for the reader.

When you use a table to emphasize and clarify information, you should place it near where it is mentioned in the text. When located on the same page as the textual material, it becomes a magnet for the eye and a hook for the reader's attention. Such a table should be brief but self-contained—devoid of unnecessary information and understood at a glance. This kind of table supports your text and identifies some of the pertinent points.

The referenced table, on the other hand, need not be short and does not have to be placed near its mention in the text. You might, instead, put it in an appendix, where it will be accessible but will not clutter the main body of your document.

When your text refers to either type of table, it should explain the purpose and should indicate when the reader is expected to consult it, e.g., "Table 2 shows this linear relation between gear wear and angular velocity," and "The complete specifications for operating speeds, capacities, and throughput are listed in Table 5 in the appendix."

Formal tables have titles and headings that make them more or less independent of the text. The informal table, on the other hand, has no title, is usually much less detailed, and is part of the text.

The main differences between formal and informal tables are illustrated by the following lists of typical functions:

91

Formal Table	Informal Table
Shows relationships among two or more sets of engineering data	Categorizes technical information
Demonstrates trends in the data	Itemizes important points
Compares information from various sources	Shows steps in a process
Lists detailed data for reference	Gives brief list of technical characteristics

Whatever its function, a table will serve its purpose only if the technical content strengthens your manuscript and if the data are arranged for quick comprehension.

Usefulness of Tables

Utilize the inherent strengths of tables. For example, tables tend to save space by providing condensed information in meaningful form. In this respect you can use a short, well-designed tabulation to replace a lengthy section of explanatory text loaded with statistics. Besides, trends and comparisons are more readily seen in rows and columns of data than in running text.

Although line charts can also show trends or comparisons, the numerical information in tables can be far more precise and easier to use for calculations.

One more advantage of tabulating information, either numerical or verbal, is that it helps you to organize your ideas in the manuscript and sometimes even to discover new relationships among the data.

Sometimes it is advantageous to *omit* a table. Tabulate information only when doing so will make the text more understandable and will not duplicate data that are in the text.

Format of Formal Tables

A table tends to be more acceptable and understandable to the typical engineer reader if its format adheres to certain conventions.[6,13] You should simplify the format, and also the content, to the point where the information can be accommodated visually in the table. Orderly, accurately labeled rows and columns are essential, as is the judicious use of white space. Avoid crowded conglomerations of data.

The main structural elements of a well-designed table, illustrated in Table 5, are:

Table 5   Structural elements of a table

			Multiple column heading		BOXHEAD
Stub heading	Column heading	Column heading			
			Subheading	Subheading	
STUB					
Line headings					FIELD
Line heading		(Tabulations of data)			
Subheading					
Subheading					
Footnotes					

1. *The title,* which includes a table number and a brief, accurate description of purpose and content.
2. *The boxhead,* which contains the vertical column headings and sub-headings.
3. *The stub,* which contains the line headings for the horizontal rows.
4. *The field,* which is the tabulation of numerical or verbal data.
5. *Rules,* which may separate certain columns or lines, or may frame the table in a box.
6. *Footnotes,* which provide explanatory comments.

Although any journal, professional society, or engineering organization may specify a special format, these six elements are universal, and each contributes to the legibility and integrity of tabular material.

The title is important. It should be readable independently of the text, but must briefly explain the content of the table that is pertinent to the manuscript. If the data are not your own, refer to the originator, as indicated in the title of Table 6. Do not repeat title information in column or line headings.

The boxhead contains headings for columns of data, which are usually the *dependent variables,* such as temperature, velocity, performance, or cost.

Table 6   Comparative properties of disk files (from Smith[5])

Disk file model	Linear bit density (Mb/cm^2)	Seek time (ms)	Geometry	
			Substrate thickness (cm)	Diameter (cm)
A	9 250	30	0.187	35
B	11 000	17	0.250	35
C	14 500	14	0.215	20
D	15 550[a]	—*	0.210	20

* Data not yet available.
[a] Estimated.

The stub contains headings for the lines of data for the *independent variables,* such as the materials, processes, or objects being measured. At the top of the stub is a heading that identifies the class of line headings under it in the left-hand column.

Because headings should be kept short, use abbreviations wherever you can. If an abbreviation is not standard, define it in a footnote.

The columns of data in the field should be grouped according to the purpose of the table, e.g., to show similarities or differences, statistical trends, or interactions among variables and parameters. When data are to be compared, it is usually better to arrange them in columns rather than in rows. Vertical comparisons are easier to read than horizontal comparisons.

Wherever possible, the columns should have a uniform degree of accuracy, each entry having only the number of significant figures that is justified by the precision. In numerical entries, fractions are usually converted to decimals, the numbers are aligned by decimal point, plus or minus signs are placed immediately to the left of the number, and a missing entry is identified by a dash:

$$542.1$$
$$-71.34$$
$$1640.$$
$$-$$

$$0.5713$$
$$-469.2$$

Rules are used sparingly in most tables; the addition of rules can be time-consuming and expensive. The usual practice is to use rules only to set off the boxhead and bottom of the table, as shown in Table 6. Vertical rules are not used except where essential to separate groups of data.

Footnotes are useful to explain why certain information is missing or to state limitations. Superscript lowercase letters are good reference symbols on numerical data. Superscript numbers are best used to indicate the literature references listed at the end of the manuscript. These two kinds of footnote symbols are illustrated in Table 6.

Chapter 15

How to Cite References Properly in Text

The Engineering Literature

Literature references serve a rather obvious purpose in any kind of paper on engineering design or development: they show what others have previously done and published. Literature references give you the opportunity to define your own innovations or improvements against that background, and they indicate how various approaches to similar engineering problems in the past differ from your own methods and results. The proper use of references, then, helps to define the novelty of your technical developments or engineering designs.

To know what has already been published in your field is, of course, a great advantage. But to find and use the references is by no means a routine matter—indeed, you face many pitfalls. Perhaps the worst of these is to assume that your colleagues know the literature thoroughly, that they know all that is going on in the field, and that you can trust them to provide you with needed references for your forthcoming paper. The experienced writer knows that second-hand references can be unreliable. The ones you see cited in other papers are not necessarily correct, and the way another author cites them or interprets them in his or her paper can be quite misleading.

The way to avoid such information gaps and traps, discussed in Chapter 11 and elsewhere,[40] is to read the literature yourself—at least the pertinent portions that you can find in a reasonable time. Searching and reading, of course, are time consuming, but after you have read one or two review articles in your field and have developed a sense of the relative importance of the various published papers, you can read selectively. In the ultimate sense, reading the articles in your field becomes insurance against writing a naive manuscript. Being sufficiently informed will save much of the time and nervous energy you would otherwise have

95

to expend in revising your paper after it is refereed. Journal referees tend to be sensitive to the importance of prior published results and to oversights in citing them and *understanding* them. The manuscript that lacks insights about previous work stamps the author as being uninformed or, worse still, as being indifferent to the important developments published elsewhere.

Attitudes About References

Literature references should not be tacked onto a manuscript as a perfunctory aid to the reader. Instead, they need to be used with taste and judgment. Although some may consider references mere "window dressing"—something to be added to a manuscript to make it look scholarly— their misuse speaks loudly for itself. Any reader quickly recognizes indiscriminate reference to the work of others. Such citations become annoying rather than illuminating and may interrupt the flow of exposition.

In this careless type of citation, the author seems to be dismissing the reference with a wave of the hand and trusting that the reader will hunt down the pertinent paper if it seems sufficiently interesting. Assume, for example, that a manuscript has a list of references at the end, including No. 14: R. F. Bond, *Glass-to-Metal Sealing Techniques,* McGraw-Hill Book Co., Inc., New York, 1976, pp. 73–79. The "hand-waving reference" might take the following form:

> The development of this widget has not been successful in the past because of the thermal mismatch[14] in the glass-to-metal seals.

From the way this book is cited, the reader cannot readily know whether Bond discusses (1) the nature of thermal mismatches in glasses bonded to metals, (2) the poor choice of materials that results in thermal mismatch, or (3) the thermal mismatch problems that occurred in the development of a particular widget. Therefore the reader must obtain a copy of Bond's book and look up the pertinent pages to find out whether it contains significant information.

The ambiguity and the waste of time can be avoided by a more specific citation:

> The development of this widget has not been successful in the past because, as noted by Bond,[14] the bonding of testalloy to Cornea Glass No. 13 results in a slight thermal mismatch and leakage in the normal range of operating temperatures.

From this example we can see how the character of the manuscript and the strength of the writing are improved by the author's attitude toward citing references precisely.

An author must remember that the voluminous technical literature grows rapidly each year and that engineering readers become swamped with the publications in their specialty. For these reasons, it is increasingly important for an author to be definitive in citing sources. A citation should indicate briefly who did what and should mention the specific development being reported.

Placement in a Sentence

The location of a citation in a sentence can, of course, have a radical effect on the meaning, as indicated by the ambiguity in the following example:

> We have examined a digital method of spread-spectrum modulation for multiple-access satellite communication and for digital mobile radio telephony.[1,2]

This kind of placement of reference numbers puts the burden on the reader to untangle the meaning either by inspecting the author's list of references or, if the titles of the papers are not adequate, by finding and reading the papers cited.

It would have been more accurate and helpful to place the reference numbers appropriately and to identify the authors in the following way:

> We have examined a digital method of spread-spectrum modulation for use with Smith's development of multiple-access satellite communication[1] and with Brown's technique of digital mobile radiotelephony.[2]

The author of an engineering paper should be precise in citing the work of others. In the example shown above, the addition of a few words and the proper placement of reference numbers clarifies the meaning by showing the relation of Smith's and Brown's work to that of the author.

The Reader's View

From the foregoing examples, it is clear that the judicious use of literature references will improve the character of any engineering manuscript, be it a journal manuscript, a magazine article, an internal report, or a student's thesis. Readers benefit because, when they read your paper and see a reference citation, they are not merely being referred to another source of information; they are also benefiting from your selection out of perhaps dozens or even hundreds of papers published in your field of work. They are being guided to what is pertinent to your particular paper and by what your taste and judgment have provided in literature references.

From the reader's viewpoint, one of the best properties of references is *completeness.* The form and accuracy of references are also important. The technique of preparing a list of references is explained, for example, in a detailed treatment by Day[20] and in recommendations for standard forms by an ANSI Standard[1] and also by Tukey[61] and O'Connor.[48] Methods of compiling a list of references on a given topic in minimum time are set forth here in Chapter 16.

Chapter 16

How to Compile
a Bibliography Quickly

The Literature Search

Gathering a list of literature references for your manuscript can be tedious but will usually reward you with new information and insights about your work. Fortunately, there are ways to minimize the time-consuming search without losing the intellectual stimulus of your hunt for information. Those methods are the subject of this chapter.

If you are to find the most literature references with the least effort, you must do two things before you write the manuscript. First, decide on the exact subject of your search and keep the scope within reasonable limits. Second, consult with your librarian about the sources and also about available information retrieval services.

You will be compiling either a bibliography of articles, books, and reports of general interest or a list of references to be cited in your text. In either case, as shown in Table 7, the time required for the search will depend on the kind of manuscript.

A lengthy or poorly conducted search of the literature can be a drain on your time and energies. Select only those references that will illuminate or support some portion of your manuscript.

Although saving time is important, do not assume that adding literature references to your manuscript is a routine job, to be disposed of as quickly as possible. The literature search is an intellectual probe, and a careful search will invariably turn up technical information that is new to you. This process tends to broaden your view of what you are reporting. The proper use of references (Chapter 15) will highlight your own contribution by showing how it differs from published papers.

Even though your manuscript may not seem to need references because it describes only the operating features of a new device or system, you should still make a brief search. If your article or report does not refer to

Table 7   Length of literature search for various types of papers

Type of manuscript	Purpose of reference list	Length of typical search
Descriptive article on device, system, material, etc.	General use: related publications for further reading	Short
R&D paper or report	Comparison of concepts and results, and of engineering problems and solutions	Medium
Survey paper, technical review, or thesis	Historical review of engineering accomplishments in the given subject	Long

published work, it will seem to be written in isolation, and your claims for originality will thus be undocumented and unconvincing. Moreover, you have an ethical obligation to give credit to original sources, or at least to cite similar published material.

A careless compilation will suffer from

- *Padding:* A detailed compilation that includes irrelevant citations does not fool anyone. Instead it detracts from your credibility.
- *Redundancy:* Duplicated references, e.g., two slightly different versions of a paper published in separate journals, are a waste of time for the diligent reader.
- *Unavailability:* Trying to locate cited books that are out of print and papers published in obscure journals is a nuisance.
- *Obscurity:* The incomplete reference (one without volume or page numbers) or the inaccurate citation is unfair to the reader.

Your search of the literature thus requires discretion and judgment. An insufficient number of references in your manuscript stamps you as being either ignorant of the engineering literature or indifferent to the work of others. Too many references suggest that you are indiscriminate and are merely attempting to impress your audience.

Whether you use manual methods or an automated retrieval system, you should learn the shortcuts to an effective search. These are discussed in the next two sections.

Manual Methods

If you do not have access to a system of electronically stored information, the first step is to consult with your librarian about the most direct and specific ways to proceed. The library may have existing bibliographies.

Or the librarian may know of journals in your field that publish review papers, and these will be a useful source of ready-made reference lists. If you find one that is reasonably recent, you might select relevant references from it.

The best way to make a manual search of the literature is to break the job down into manageable pieces. A good strategy is to start with a small segment: *a working bibliography.* Compile a short, selective list of books, articles, and reports dealing specifically with your topic. After examining this compilation, you can either search further and expand the list or condense it to suit your needs. The length of your final listing will be limited by (1) the amount of detail that seems appropriate and (2) the time you can devote to the search.

As an example of the need for a quick, preliminary search, we can use the paper on the design of a solar energy system in Chapter 4. The outline provides a detailed list of possible topics, but the main subjects are the solar radiation collector array and the heat transfer in the circulation loop. These two essential topics, without the peripheral details in the outline, can define the limits of the search.

Start with a library search for book titles, which you can find in the subject catalog under the heading SOLAR ENERGY. You also may find it convenient to consult the standard reference *Books in Print,* published by R. R. Bowker Company, New York, under the heading SOLAR HEAT-ING. Select only the titles that deal with design methods. Then proceed to search the periodical literature, which is more detailed and more current than books and monographs.

The best guides to the periodical literature of engineering are the abstracting and indexing journals. Some cover broad fields, such as *The Engineering Index, Science and Technology Index, Chemical Abstracts, Government Reports Annual Index* (NTIS), and *Scientific and Technical Aerospace Reports.* Others are more specialized, like *Computer and Control Abstracts, Electrical and Electronics Abstracts, Metals Abstracts, Geological Abstracts, Applied Mechanics Reviews.*

Suppose you were doing a search for the solar energy paper. You could choose a comprehensive service like *The Engineering Index* because of its extensive coverage of engineering design papers. The abstracts in the annual volumes are listed according to subject.

A glance at the subject categories and their cross references in *The Engineering Index* shows two that relate immediately to the chief topics in our paper: *"SOLAR RADIATION–Concentrators,"* and *"SOLAR POWER PLANTS."* Figure 15 is a portion of a 1982 page in the *Index* containing abstracts on solar power. The abstract for a given paper helps you determine quickly whether the reference is suitable for your compilation. If you are to include internal technical reports in the bibliography, your company or school librarian can show you how to search for them.

007415 PERFORMANCE STUDY OF FORCED CIRCULATION SOLAR WATER HEATERS USING PACKED-BED COLLECTORS. A performance study of forced circulation solar water heaters using packed-bed solar collectors is presented. Iron chips, gravel and stones have been used as packing materials. Thermal energy stored in the tank, system overall efficiency and pay-back capital for these solar water heaters are compared with those for solar water heaters using a plane collector. It is observed that the performance of the solar water heater is improved appreciably by packing its collector with packing material. A solar water heater using an iron chip, packed-bed collector shows the best performance. 7 refs.

Mishra, C.B. (Birla Inst of Technol, Bihar, India); Bhat, A.K. *Energy Convers Manage* v 21 n 2 1981 p 121-123.

007416 EFFECT OF ADIABATIC CO-PLANAR EXTENSION SURFACES ON WIND-RELATED SOLAR-COLLECTOR HEAT TRANSFER COEFFICIENTS. The heat transfer response to framing the thermally active cover surface of a flat plate solar collector with adiabatic coplanar extension surfaces has been investigated by wind tunnel experiments. Various framing patterns were employed (leading edge and/or trailing edge and/or side edge framing), along with frames of different width. The experiments were performed for various angles of inclination of the plate surface relative to the oncoming airstream and for a range of Reynolds numbers. It was found that the wind-related heat transfer coefficients can be substantially lower when the collector is framed than when it is unframed. An estimate of the possible reduction of the average heat transfer coefficient can be obtained from the equation $h/h^* = (L_c/L_f)^{1/2}$, where h and h^* respectively denote the coefficients in the presence and in the absence of the frame. The quantity L_c is a dimension that is characteristic of the thermally active area of the cover surface, while L_f is a characteristic dimension of the outer edges of the frame. 5 refs.

Sparrow, E.M. (Univ of Minn, Minneapolis, USA); Lau, S.C. *J Heat Transfer Trans ASME* v 103 n 2 May 1981 p 268-271.

Concentrators See Also SOLAR POWER PLANTS; SPACECRAFT—Design.

007417 HIGH TEMPERATURE SOLAR CONCENTRATOR DESIGN. An optical sub-system design for a solar concentrator having applications to high temperatures thermal photovoltaic concepts has been developed. The baseline system consists of a spherically segmented primary mirror optically coupled to a compound parabolic concentrator and a hemispherical receiver. Secondary concentrators having conical, parabolic and hyperbolic cross sections are considered parametrically. Performance trade-offs relative to, concentration ratio, system configuration and manufacturing sensitivities have been evaluated using geometric ray tracing techniques. The advantages of using a two-stage concentrator over a single collecting aperture are discussed. Performance of typical systems are reviewed covering transmission, flux distribution, and tolerance sensitivities. 4 refs.

Wientzen, Richard (Itek Corp, Lexington, Mass, USA); Davis, W.J.; Forkey, R.E. *Proc Soc Photo Opt Instrum Eng* v 237, Int Lens Des Conf, Proc of Tech Pap, Mills Coll, Oakland, Calif, May 31-Jun 4 1980. Publ by SPIE, Bellingham, Wash, 1980 p 281-291.

Figure 15 Abstracts on solar power plants selected from the January 1982 issue of *The Engineering Index*.

BIBLIOGRAPHY

Avezov, R. R., "Investigation of Heat Transfer and Efficiency of Screen for Tubular Heat Collectors of Low-Molecular Weight Solar Water Heaters," *Appl. Sol. Energy* 15, No. 1, 27–30 (1979).

Azkmov, S. A., "Calculation of the Optical Characteristics of High-Power Two-Mirror Solar Furnaces," *Appl. Sol. Energy* 15, No. 2, 23–29 (1979).

Demichelis, F., "Concentrator for Solar Air Heater," *Appl. Phys.* 19, No. 3, 265–268 (July 1979).

Grossman, G., "Development of a Spherical Reflector/Tracking Absorber Solar Energy Collector," *Israel J. Technol.* 17, No. 1, 5–11 (1979).

Kudrin, O. I. and Abdurakhmanov, A., "Selective Radiation Absorption as a Means of Improving Efficiency of a High-Temperature Solar Power Plant," *Appl. Sol. Energy* 15, No. 4, 36–43 (1979).

Lebens, R. M., *Passive Solar Heating Design*, Halsted Press, New York, 1980, p. 220.

Moore, R. R., "Optimized Grid Design for a Sun-Concentrator Solar Cell," *RCA Rev.* 40, 140–152 (June 1979).

Pillai, P. K. C. and Agarwal, R. A., "Black-Liquid Solar Collector," *Sunworld* 3, No. 4, 108–110 (1979).

Scheller, W. G., *Solar Heating*, Bobbs-Merrill, Inc., Indianapolis, 1980, p. 175.

Singh, R. N., Mathur, S. S., and Kandpal, T. C., "Using a Fin with a Parabolic Concentrator," *Int. J. Energy Res.* 3, No. 4, 393–395 (Oct.–Dec. 1971).

Sobirov, O. Yu., "Foam-Film Parabolic-Cylindrical Solar-Energy Concentrator," *Appl. Sol. Energy* 15, No. 1, 25–26 (1979).

Tracey, T. R., "Low-Cost Central Receiver Solar Power Plant Using Molten Salt as a Heat Transfer and Storage Medium," *Energy Technology Proceedings*, Energy Technology Conference 6th, Washington, D.C., February 26–28, 1979, Vol. 2, pp. 1059–1065.

Wen, L. and Huang, L., "Comparative Study of Solar Optics for Paraboloidal Concentrators," *ASME*, Paper No. 79-WA/Sol-8, for December 2–7 Meeting, p. 13 (1979).

Zakhidov, R. A., Abdurakhmanov, A., and Klychev, Sh. I., "Optimal Geometric Parameters of Cavity Solar Collectors with Selective Radiation-Absorption Properties," *App. Sol. Energy* 15, No. 1, 13–15 (1979).

Figure 16 Working bibliography of recent references on solar collectors and heat transfer studies in system design (compiled from a manual search).

Then prepare a preliminary list of books and papers, like the one in Figure 16. Retain on your list only those that are really pertinent. If you feel that what remains needs to be augmented, dig further in the card catalog and the periodical indexes, going back as many years as is appropriate.

Table 8   Some commercial search services having on-line access to databases on engineering papers and reports*

Name of service	Organization
BRS	Bibliographic Retrieval Service, Inc., Scotia, N.Y.
DIALOG	Dialog Information Services, Inc., Palo Alto, Calif.
ORBIT	System Development Corporation, Santa Monica, Calif.

* Some databases cover, in addition, the literature of science, education, business, etc.

For accepted abbreviations of professional and industrial engineering periodicals, see *PIE, Publications Indexed for Engineering,* published by Engineering Information, 345 E. 47 St., New York, NY 10017. More than 2,000 such journals and magazines are listed.

The Automated Search

A computer terminal that gives you access to an information retrieval system is a powerful tool for searching the engineering literature. Although modern equipment can complete a search through many thousands of documents in an incredibly short time, the quality of what you retrieve will depend very much on the nature of your inquiry.

The machine search is a flexible process in which the computer looks for key words or descriptors in its database to match your requested topics but also has other options. Many systems make a "full text search," looking for the requested key words and phrases in each title and abstract instead of being limited to a search of general subject groupings of the documents. In addition, retrieval systems use Boolean logic (AND, OR, NOT, etc.) to permit you to request combinations of certain topics and exclusion of others, and to specify the conditions of search, such as time interval and numerical limits.

Another important aspect of information retrieval is the selection of the best database for your particular query. Your librarian can choose from vendors who have access to many separate data banks. A few commercial services are listed in Table 8. Individual data banks have also been set up by government and private industry. Some have specialized coverage, e.g., patents or Ph.D. dissertations. Other have broad coverage of journal papers, books, and industrial and government reports. Table 9 lists a few representative databases that exist at this writing.

Your librarian or information specialist will help you formulate your query in a language that will extract information from the computer files. A great deal depends on how carefully you make the request. You must provide a list of key words and phrases. Supply modifiers, restrictions, and limitations. Do not confine your request to a single term without qualifications. For example, if you ask for a search on "very large scale inte-

Table 9   Some representative databases for engineering literature

Name	Coverage	Producer
CA SEARCH	Content of *Chemical Abstracts:* chemistry and other sciences, plus some engineering disciplines. Papers, patents, and books	Chemical Abstracts Service
COMPENDEX	Content of *The Engineering Index:* journal and conference proceedings papers	Engineering Information, Inc.
COMPUTER AND CONTROL ABSTRACTS	Engineering and scientific papers	Institution of Electrical Engineers, London
DISSERTATION ABSTRACTS	Many disciplines: doctoral dissertations	Xerox University Microfilms
ELECTRICAL & ELECTRONICS ABSTRACTS	Journal and proceedings papers, electrical and electronics engineering	Institution of Electrical Engineers, London
ENERGY NE	Scientific, engineering, and nontechnical aspects of energy	Environment Information Center, Inc.
GPO MONTHLY CATALOG	Citations from all U.S. Federal Government publications	U.S. Government Printing Office
METADEX	Science and engineering of metallurgy	American Society for Metals
NTIS	National Technical Information Service	U.S. Dept. of Commerce, Washington, D.C.
SCISEARCH	Many disciplines in science, engineering, and technology: journal papers and books	Institute for Scientific Information
U.S. PATENTS	Patents registered in the U.S. Patent Office	Pergamon International Information Corp.

gration" you will undoubtedly receive a list of hundreds of articles on this kind of microcircuit. The bulk of material you will have to scan and select would then defeat your purpose—to compile a bibliography *quickly*.

Your query should supply some or all of the following information:

- *One or more technical concepts* (key words or phrases, spelled correctly)
- *Related terms*
- *A combination of two or more concepts* in a single document
- *Limitations:*
  —Confining the search to a portion of the specialty
  —Cutting off the search at a specified time or limiting it within numerical bounds
  —Broadening the search to additional portions of the specialty

• *Alternatives to subject searching:*
  —Finding citations by a given author or authors
  —Finding citations originating at a given author's affiliation
  —Finding existing bibliographies

In our example you might make the following request:

> Make a search on *design of solar energy systems* but only on the following specific subjects:
>
> Solar energy collectors*
> Solar energy concentrators*
> Solar power plants
> Transport fluids**
> Heat transfer in circulation loops
> Energy gain
> Solar collector arrays
>
> Confine search to journal papers in 1984 and 1985.
>
> * Search for collectors and concentrators only in solar cell arrays, not in photovoltaic cell arrays.
> ** Search for transport fluids only in solar energy systems.

These query specifications will help to prevent the sytem from selecting irrelevant references.

Manual Versus Machine Searching

In general, the manual search is better for simple topics on which there is a small body of literature. The automated search is usually better for more complex topics and a large volume of publications.

In addition, manual and automated searches have different kinds of flexibility. When you look through the indexes in a library, and perhaps browse through journals, you tend to be sidetracked by information that may be useful but is outside your interests. An automated search gives you flexibility to use complex criteria when formulating the query. Nevertheless, no matter how carefully you phrase your query, you will usually have to edit the printout to eliminate abstracts that are not pertinent. All authors must, to some extent, be editors. And they need to use all their editorial skills to perfect their own manuscripts, a process which is the subject of Chapter 19.

Chapter 17

How to Write and Publish
a Thesis or Dissertation

Preparing the thesis required for an advanced degree can seem like a mere ritual—and a necessary evil—to provide you with proper credentials for graduation. For a master's thesis you might feel this way if you are not really excited about the selected topic. Although writing a doctor's thesis can be more of an intellectual challenge, it tends to be a tedious job, and any doctoral candidate experiences uncertainties and frustrations before it is finally accepted by the thesis committee.

But in either case the planning and writing can be an opportunity, not only to demonstrate scholarly competence for the thesis committee but also to make a contribution to the engineering literature. Indeed even a master's thesis, after publication, can turn out to be a springboard to a future career.[31]

Unfortunately many theses representing original research—both master's and doctor's—now reside in dusty retirement on the university library bookshelf because the student had lacked a proper plan and motivation to publish in the journals. For many degree candidates the rigors of getting acceptance by the thesis committee can dampen any enthusiasm for finding a publisher. For others a lack of sensitivity in the thesis to the needs of engineering journals can spoil the chances for publication.

Avoid these situations! Seize your opportunities at the start of the project by planning your thesis to be a professional achievement and a bid for peer recognition in your field.

Whether or not you are *required* to publish, you need to develop a dual strategy for such a plan. Select a thesis subject that will not only satisfy your thesis advisor but will also be publishable in a journal of your choice. You may even plan to write and publish the theoretical portion of your project as soon as you have developed the theory. After confirming your hypothesis by testing, you could then publish a companion paper

107

verifying the theory. Thus two journal papers can evolve from your thesis—sometimes referred to as a "dissertation."

Thesis or Dissertation?

In recent times the original meanings of "thesis" and "dissertation" have become somewhat blurred. In academe you may hear these terms used interchangeably. The master's requirement is often called a thesis and the doctor's a dissertation. Regardless of these confusions, observing the fine distinctions of meaning will clarify the target of your manuscript and perhaps show you how to hit a bull's-eye.

Theses and dissertations consist of an original inquiry into a significant problem. Traditionally the main difference between the two has been the form of the inquiry, defined as follows:

A *thesis* adopts a specific view, defends it, uses logical argument, and offers a valid solution.

A *dissertation,* as formally defined,[31] tends to be less rigidly constructed. It is a systematic, in-depth discussion of a subject in the form of a learned discourse.

Even though the degree requirement in any engineering or physical science curriculum is usually fulfilled by the *thesis* form as defined above, your advisor may loosely refer to your manuscript as a "dissertation."

Requirements for Engineering Degrees

The master's degree and the doctorate have several goals in common:

- Competence in scholarly research
- Originality and inventiveness
- A demonstration of understanding of an engineering problem
- Insight into the solution of the problem
- Clear, orderly presentation of the concepts and results
- A transition from the world of facts into the realm of ideas and reasoning

In other respects the requirements are somewhat different.

The Master's Thesis

One of the main purposes of the master's thesis is to demonstrate the mastery of methodology and to show ingenuity and independent thinking.

But this need not be a mere exercise, to be duly submitted to the thesis committee for a stamp of approval! If a good topic is chosen (as discussed in the next section), the thesis can and should be an original contribution, written on a professional level and eventually made available to the engineering community.

You may not be required to do experimental work for a master's thesis, dependin' on the ground rules at your school. A critical survey of the literature on a special topic can be an original contribution, both for fulfillment of degree requirements and for publication. In this kind of thesis the candidate compiles and analyzes various published papers on the given topic. The critique includes an identification of the significant research published so far, a commentary on unresolved questions and discrepancies, and new insights and discussion.

The Ph.D. Dissertation

The doctor's degree is more demanding in the following ways. The dissertation should offer a research accomplishment, with evidence that the candidate is able to do reliable intellectual work alone. Compared to a master's thesis, the examining committee will expect to see evidence of higher analytic skills and more sophisticated reasoning. Most important, aside from originality of concept, is the ability to design and control the research and experimentation.

In addition to demonstrating a solid grasp of methodology, the Ph.D. candidate must provide sound and significant conclusions to the research study, i.e., the author's "contribution."

Selecting a Topic

For either the master's degree or the doctorate, much depends on your selection of a good topic. First, you must choose one that will permit you to demonstrate original thinking and satisfy the thesis committee. Second, your topic should be publishable in a journal so that you will become widely known for your work. Third, the research can actually be your entry into an engineering specialty and the start of your professional career.

From your viewpoint as a degree candidate, the first of these three goals—getting the thesis approved and accepted—will tend to swamp out the other two. When facing the agony of choosing the *right* topic, the question of getting the topic approved by your thesis advisor, and the job of doing the research and writing the manuscript, should you really consider such remote events as journal publication or starting a professional career?

Here is a direct answer to that complicated question. Yes, you should consider tackling all three goals at once because the extra time spent will be a very small percentage of the total effort to write the thesis! And the personal gains can be enormous.

The extra effort consists mainly in picking the right topic. And just how can you accomplish this miracle of killing three birds with one stone? Indeed, the three goals are not always compatible. For example, a thesis of academic interest that would bring high praise and acceptance by your committee might not be at all suitable for the readership of a professional journal. And a methodical, logical, and exhaustive treatment of your engineering subject, admired by the faculty, will not necessarily launch you into a career.

Fortunately, there are specific ways to pick a thesis subject that is compatible with all three goals. Finding such a subject will always be done in close consultation with your faculty advisor. You may or may not be asked to choose a topic from several that would support your advisor's research program. In such a case the faculty member can help you modify one of these topics to improve the chances for publication. Or you may have full freedom to choose *any* topic, which may or may not be provided by a funded grant. In any case, selecting a topic has to be a new approach for you because the research for the thesis or dissertation will be quite separate from your course work. It is beginning a new program and will probably be the first large independent work in your career.

When looking for a topic consider the following:

1. The subject must have the potential for demonstating original thinking.
2. It has to be researchable, involving a concept that can be called a contribution to your field.
3. The work should be of a size that you can complete within a reasonable time as a thesis project.
4. Your subject should not depend wholly on the support of a thesis advisor who could conceivably lose position in the program and not serve on the acceptance committee when you finish the manuscript.[55]
5. Take care to avoid those engineering problems that demand ingenious solutions but are of no real interest to the world of engineering outside the university.
6. A good topic is one of mild technical controversy and one that has not been fully explored. Of course, you should avoid topics that conflict with the research results of professors on your thesis committee.
7. Look particularly for gaps or missing links in the papers in current issues of your favorite engineering journals. Plan to do an illuminating piece of research *that will fill a need for the readership.*
8. Be sure to tackle a technical problem that interests you. If you can pursue it with enthusiasm, you are on the right track. Don't start your

thesis work until you have found an engineering problem that intrigues you and begs for a solution or clarification.

9. Most important, your topic must be timely and technically significant to your (future) peers who are now working in this specialty.

These nine items suggest how the choice of a topic can help you toward the three goals. The first four on the list aim toward future acceptance of your thesis by the committee. The last five aim at future publication and at a career as an engineering graduate.

Items 1 and 9 deserve special attention. As a degree candidate you are expected to contribute something to the state of the engineering art. Later on, as a graduate working in your chosen field, your subsequent contributions will probably overshadow your research for the thesis or dissertation. Even so, the topic chosen by any degree candidate must be significant. The following example will show how you can proceed.

An Illustrative Example

Our hypothetical example will be that of a degree candidate who finally chose the solar energy topic discussed in other chapters of this book. The outline for that paper, "Design of a High-Efficiency Solar Energy System," appears on page 25. Our candidate, working toward a degree in Mechanical Engineering, went through the usual procedure of discussing potential topics with his thesis advisor and other professors. Various un-solved problems in mechanical engineering were suggested. In further search of a likely subject the student, whose main interests were in heat transfer problems, browsed the journals of engineering and physics. He also did a literature search on solar energy and heat transfer. He noticed that papers on solar energy schemes were usually concerned with methods of increasing efficiency and that there were several competitive avenues of research. He also observed that the general topic had wide appeal because of the potential for fuel economy in heating systems.

The student's strategy developed along the following lines. Instead of attempting to design an entire system, he decided on a research project of a more manageable size. The purpose of the research would be to improve the efficiency of two small components of the solar energy system. These components would be a newly designed light collector and an optimum composition for the transfer fluid. The combined effect of the two im-provements would then give an impressive research result for the whole system. The student's strategic approach had to be reinforced, of course, by feedback from faculty advisors.

This strategy thus set up a three-pronged goal for the thesis. The research project would be attractive to the approval committee, and the finished thesis should provide good material for a journal paper, which later on would enhance the professional reputation—and the career potential—of the new graduate.

Selecting a topic in this way can be far more productive than the more conventional method of placing future publication at the bottom of your priority list.

After you have selected a topic the next step is to research the subject further and draw up a proposal.

Preparing the Proposal

A proposal or prospectus to be submitted to your faculty committee is intended to clearly define your thesis program and to show the rationale for the research. The approving committee can then examine your proposal for appropriateness and validity.

The proposal also has other purposes. One is to establish the tone and pattern of relationships between you and the faculty. Another is to provide a framework of your plan so that the professors who monitor the program can advise you about your progress.

The form of your proposal will depend on the thesis advisor's requirements. For a master's degree they may consist of an extended statement of the problem and an outline of the proposed manuscript, indicating the methods to be pursued and the kind of research results to be expected. Some master's programs will require more detail.

A proposal for a Ph.D. thesis has to be lengthier and more formally constructed, taking the form of a "mini-thesis." The longer and more detailed it is the better. This document should be as much as half the size of the finished thesis. Although faculty advisors might not want to be burdened with voluminous details, a thoroughly prepared proposal is to your advantage. It makes a clear statement of your case for the review committee. The proposal also gives you the opportunity to resolve the committee's demands and criticisms at this preliminary stage of your thesis. Once the committee members have reviewed the details, their approval of the proposal is a kind of "contract," as suggested by Sternberg,[55] that extends to the finished thesis itself.

There is no standard format for the sections of a thesis proposal. The general form is usually specified by your department. The structure could be as simple as Abstract, Introduction, Theory, Data, and Discussion. The following structure, however, is fairly conventional:

The Problem
The engineering problem is stated in meticulous detail, including the reasons for attempting to solve it.

The Hypothesis

The rationale for solving the problem is described in depth.

Review of the Literature

This is one of the more important sections. If it is skimpy it breeds suspicion of inadequate penetration into the technical subject and its history. The list of cited references should be lengthy but pertinent, and not padded with unrelated papers.

Methodology

This section traces the thread of research: the procedures and methods for obtaining the data or developing the design.

Statistical Design

The analytical tools and the ways data are to be handled are shown here.

Data

The anticipated numerical results and graphical configurations are shown sketchily. These will eventually be the basis for the conclusions.

Discussion

Your initial attempt at a commentary gives the results of the research, including its relation to previous published work, the difficulties and unresolved questions, and limitations of the technique and the data. (This section could be combined with your Conclusions.)

Conclusions

The validation of the engineering solution, its significance, and applications of results.

Appendices

A reference section giving derivations and other supporting details that will not be essential to the Discussion and Conclusions.

Bibliography

The list of prior published papers and books, the result of extensive search of the literature.

After your proposal is accepted start right away, while your enthusiasm is still high, on the research and the manuscript.

Writing the Manuscript

Some portions of the proposal can be used almost verbatim in your final manuscript. The Abstract will probably contain all you will have to offer in the finished version. The Introduction will probably be shorter than in the thesis, as illustrated by the examples in Chapter 7. And since you have already developed the theory and methodology for the proposal, you can incorporate most of those sections in the thesis.

In preparing the rest of the manuscript you are expected to rely on the skills in research, analysis, and proof that you have developed in writing term papers. But in most respects, writing the thesis differs from completing your course assignments. You are now on your own, and the main guidance from faculty advisors will consist in reading and criticizing the drafts of your chapters.

In constructing those chapters you need to follow the format requirements in graduate student handbooks[55,58,62] and the instructions in department seminars or style sheets. The best models for structure and content are *published papers in the engineering journals.* The more you read in the course of your literature searches the better you will be equipped to write professionally. Writing your thesis, however, involves more than following instructions and observing good models. Structuring and preparing the thesis is brainwork and is the end product of your scholarship and intellectual curiosity.

Your own attitude toward the thesis work will inevitably be reflected in the writing. If the research is dull and boring to you, you will write a dull, boring manuscript. How, then, do you write an interesting, readable thesis? You can generate interest and readability, not by attempting to force it, but by developing it in the following way.

After you have searched the literature on your thesis subject, look over your outline, like the one on page 25. Identify those items that have not been previously known and understood. Examples are "I.C. New design approaches," "II.A. Principle of multistage concentration of light rays," "V.A.2. Transport fluid with high heat capacity," and so on. These are your contributions to the state of the art of designing energy systems. When you have identified such portions of your outline, you can liven your manuscript by emphasizing them. Plan to devote more space and more explanation to these important points and to de-emphasize other more routine sections by compressing their length. Do the same with figures and tables. Use them to illustrate and emphasize the significant parts of the thesis and resist the temptation to include volumes of data that do not contribute anything important to your discussion and conclusions. These simple methods of proportioning the manuscript will always add reader interest.

This chapter has shown how to plan your thesis according to degree requirements and how to select a topic *according to your own requirements,* e.g., publication and professional recognition. When at last you have finished the thesis manuscript and submitted it for approval, you will undoubtedly want to rest and await the outcome. This is the time when your frustrations and struggles have yielded a finished product and your manuscript is a personal triumph, awaiting the committee's imprimatur. Especially if it is a dissertation you will have to defend it when the time comes. But during this waiting period, while you are still warm on the subject, is the time to plan your journal manuscript. It will not require much planning, but if you delay now may never get around to it.

Look once more at the outline. What can you delete to convert it into a journal paper? Certain items will be obviously of little interest to the readership: portions of the extensive historical review so necessary for a thesis; mathematical derivations intended to demonstrate competence for the committee; details of methodology that are already well known; figures and tables that are peripheral to your main theme; and the expansive bibliography, which usually can be cut down to a lesser size for purposes of a journal paper.

Because at the start you had also laid the groundwork for a journal paper, reducing it to a suitable length is mostly a mechanical job. Get it retyped and submit it to the journal editor before your enthusiasm cools!

The ordeal of writing an engineering thesis and getting approval is the experience of a lifetime. If your career will be in research and development work, writing a thesis is your entry into an intellectual tradition. And the two-pronged approach—writing reports for your engineering organization and publishing in the journals—will become part of your professional development.

Chapter 18

How to Publish Your
"Confidential" Results

Research and development work can be frustrating if security restrictions prevent publication of your results in the journals. In *company work,* your engineering development of a new product may include proprietary information. In *defense work,* details of a new weapon or a novel design technique may have to be withheld to protect national security. In some cases, even unclassified data must be kept under wraps because of government policy. However, when you are working in such a "can't publish" atmosphere, there are several initiatives you can take to make your technical achievements known to the engineering community.

My purpose is to suggest some legitimate steps that will start you on the road to eventual publication. Of course, you cannot publish company secrets or compromise national security. But you can devise ways to publish professionally at an opportune time, with proper clearance, and become known in the engineering literature for your contributions to the state of the art.

Even if writing a journal paper seems completely out of the question, certain strategic choices will still be open to you. These choices are the subject of this chapter.

Although information security is quite important in business and government, there will always be a point of limiting returns. After the restrictions against publication have outlived their usefulness, they can become a bottleneck instead of a protection.

The Information Bottleneck

In high-technology areas, the Information Age has a special significance. Inevitably, the rapid pace of new technical developments, both industrial and military, requires a security lid on certain types of data.

116

Without controls on the engineering information the competing forces at play tend to get out of hand.

Industrial R&D. Marketing strategies demand that unique engineering know-how be withheld from commercial competitors. Proprietary information includes new designs and procedures that are not yet protected by patents, special production techniques that solve manufacturing problems, and future plans for developing and marketing the company's products. Divulging the company's key assets not only affects sales and profits but, in the long run, also endangers job security.

Military development work. Defense strategies demand protection of engineering information on weapons development. Security leaks to potential enemies ultimately weaken national defense and may endanger lives and property.

Technical developments are the key to military superiority and, of course, are essential to higher sales and profits in today's technological industries. But these developments can be hampered by overprotection and the resulting "information bottlenecks." Whenever security restrictions *can* be lifted and your work published in the outside literature, the following advantages become evident:

- The normal flow of information, in journals and professional society conferences, will resume. This becomes a two-way flow between you and your peers: publication of results, audience reaction, comment, and discussion.
- In the case of defense contract work, "technology transfer" permits industry to pick up declassified engineering developments for use in commercial products and systems.
- Outside publication becomes a stimulus to engineering effort, as discussed at the end of Chapter 5.
- You gain technical visibility, your company and/or the government contractor gains in reputation for new technologies, and your marketing division gains support.

Security restrictions, however, are not always automatically lifted, either in industrial developments or in defense contract work. Technical reports that are "Company Confidential" can remain in that status after the classification is no longer necessary. Declassification procedures are standard in every security system but are frequently ignored through oversight or indifference. Devoting time and effort to declassifying documents seldom has high priority in an engineering department's activities. Instead, there are always new projects to start and a continuing succession of engineering problems to solve.

The same situation is often true in defense work. Confidential and Secret documents can remain classified long after there is any reason to secure the information.

But one fine day, months or years after you have finished a classified piece of engineering work, your manager might suddenly inform you that your data can now be released and it might be a good idea to "write it up for publication." At that late date, however, the prospect of several successive lead times can be forbidding. Many months can elapse after the manuscript is written, cleared for release, submitted to a journal, sent to referees by the editor, revised, prepared for the printer, proofread, and finally printed and published. When such a delayed paper finally reaches readers it can be "old hat" and no longer of interest.

For these reasons you should take steps early in the development cycle to break the "information bottleneck" by planning for eventual declassification and journal publication.

Breaking the Bottleneck

Although your department manager may want to help you get published, it is not always easy to set up a publication plan before the classified work is finished. You could meet resistance at first, but your game plan will succeed if you devise the right strategies.

In the normal situation your department manager will be struggling to use the available resources in order to get the engineering problems solved, to stay within the budget, and to finish the project in the allotted time. During this busy period, planning for eventual publication might not even be a visible objective, especially if you are working under a security umbrella. In such an atmosphere of brisk project activity the best approach is to remind your manager that planning future publication has certain important advantages for the department. For this discussion with your manager be sure that you come equipped with a topical outline for your proposed paper. It should highlight your technical achievement.

Point out that your engineering company is competing in the marketplace and, in view of competitive developments in other companies, needs to display its technological achievements. Early publication of your paper at the end of the project will enhance your department's technical reputation.

Or, if you are working under a defense contract, you might point out that journal publication at project end becomes a public record of engineering competence in your department—an aid to gaining new contracts and to gaining introductions to government laboratories and to researchers you might not otherwise expect to meet. Positive approaches of this kind can spur your management to get your future manuscript declassified promptly. More specifically, you can also take any of the following four steps to expedite matters.

Request a Release Date

A good strategy is to request from your manager an estimate of the first possible date for release of the details of your work when it can fit into the marketing plans. This gives you the opportunity to work toward a date for completing your manuscript and sending it to a journal.

In development of a new industrial product or system, the date of first shipment to customers is usually a time for releasing at least some of the details of the engineering development. Manufacturing techniques might have to remain confidential for a while, but the final design and configuration of the product can be readily determined by anyone who purchases the item, takes it apart, and analyzes it. Therefore, after the initial shipment you should be able to publish your paper or present it at an engineering conference.

In defense work, clearance of your paper may be more complicated. Sometimes the contract specifies release and publication of technical details at completion of the contract. But usually any manuscript for outside publication must be sent for security clearance through the contract representative.

Make early inquiries about an *anticipated* release date for specific areas of technological detail. If the contract representative does not readily specify a date, try again. Your inquiry is legitimate. It may require some probing and some decision making by the contractor, but it does deserve an answer.

In either case—industrial work or defense contract—once you have an estimated release date you can set up a publication plan. But if you can't get such an estimate there are other avenues of approach.

Plan to Publish, Regardless

You are certainly at a disadvantage when you have no estimate of when your work can be declassified, but you can still make a plan. Follow my suggestion for writing in increments (Chapter 5). Prepare an initial outline and write portions of the manuscript at intervals as the work proceeds.

The incremental method of writing gives you three strong psychological advantages. First, you can tackle the writing job with enthusiasm because you have established a goal, i.e., a completed manuscript at the end of project work. Second, by preparing a manuscript, you will have taken the initiative toward eventual declassification. Third, the insights you derive from organizing and writing give you a better understanding of technical matters and thus actually help the work.

The resulting security check for technical content of your classified draft may require that you make deletions, in which case your gambit has been immediately successful and you can revise, clear, and publish the paper. If, on the other hand, your paper has so much privileged data that it cannot be released in any form at this time, your gambit has not really failed. Nothing stays classified forever. The time will come when declassification by the proper authority is finally feasible. You will then have a decided advantage. Instead of starting the planning and writing from scratch, you will have a finished paper, ready to submit to a journal.

Omit Motivation or Applications

One way to help get your work published is to delete from the manuscript certain portions that would show the motivation of your paper. Then you can request clearance for release of the deleted version. This does not mean that you need to omit all the "goodies" from your manuscript. Fortunately, many such papers with minor deletions are successfully published each year because they were acceptable to journal editors and valuable to readers.

Omitting important information to make your writing acceptable is not a standard technique of exposition. Although such methods are seldom, if ever, taught in English class, they are a valid way of stimulating information flow in the engineering and scientific communities. These techniques for writing an acceptable paper are important both for industrial R&D and for defense work, especially research projects.

You will need to ask some probing questions of your management, the marketing people, or your contract representative. Find out which technical areas are sensitive and which ones are not. Frequently you will find that either Company Confidential or defense Secret or Confidential results can be published if you describe the basic research or development but do not reveal the motivation for the work or its potential applications.

For example, you might be permitted to write an electronics paper on novel circuits for dual digital conversion at microwave frequencies without explaining how it is used in military radar equipment. Or you may be able to publish the theoretical basis for producing ultrafine lines in photographic films by X-ray bombardment without explaining the complexities of applying this method to produce microcircuits for computers.

Omit Critical Data

In some circumstances you will be permitted to show the motivation for your work and its applications but can publish only if you omit certain kinds of critical data, such as the times and temperatures of a manufacturing process. Publishability will then depend on the technical content remaining in such a manuscript. The omissions could ruin chances for acceptance by any journal. But not necessarily. If the content still contributes something new and interesting to the literature, submit it for publication. It may be accepted regardless of the omitted data. Many papers are.

Chapter 19

How to Edit Your Manuscript
for Content and Integrity

The Author as Editor

In writing any manuscript you become your own editor every time you change a word, add a phrase, or reorganize a paragraph. You might think of this kind of editing as an application of your engineering training to clarify problems and to arrange the findings in logical order.

After you finish the draft, your manuscript will be subject to further editing. If you work for an engineering organization, the draft may go to an editor for copy editing. If it is to be submitted as an article for publication, it will also go to a journal editor, who may modify it further.

As an engineer author, your initial self-editing differs, in certain important ways, from the work of a copy editor. For this reason you should not turn over to an editor a poorly organized, hastily written draft. Your responsibilities for the technical adequacy of the manuscript include several aspects that you cannot delegate to an editor:

- *Completeness and accuracy of data.* An editor cannot always identify missing or wrong information, e.g., numbers you had omitted from a tabulation or data you had miscopied from your engineering notebook.
- *Clarity of the basic concept.* If you are the father (or mother) of the thought, an editor cannot be expected to supply the details of your thinking.
- *Adequacy of the engineering solution.* Someone else cannot fill the gaps in a sketchy explanation of how you solved the technical problem.
- *The main thrust of your manuscript.* Only you can decide which portion should receive the most emphasis: theory, design methods, engineering properties, applications, etc.
- *Technical interpretation of the results.* In your manuscript any analysis and commentary about your findings must reflect your own thinking.

If you have the services of a technical editor in your organization, you will get help in matters of grammar, mechanics, format, and style. If you do not have the aid of an editor, you assume the responsibility of copy editing yourself.

The following two sections show how to edit for content, which is your best insurance against possible future barbs and criticisms of a journal referee, and how to edit for clarity and integrity, which is the essential refinement of any manuscript.

Editing for Content

It is not easy to analyze your own manuscript for technical content. The best way is to stay away from it for a few days and then to have a fresh look. Divest yourself of pride of accomplishment and of your euphoric attitude about the work and the document. Then you will develop the psychological set needed to exercise your critical faculty.

If you measure your paper by the kind of yardstick used by journal referees, you gain a new perspective of the validity and effectiveness of your writing. You can use the box on the next page as a yardstick. It served for many years as my instructions to journal referees. Fortunately, the criteria listed in the box apply generally to all kinds of engineering manuscripts, including R&D papers, trade magazine articles, engineering theses, and technical reports.

If your manuscript is a recommendation report or a proposal, written for a specific audience, add the following questions to the list:

Does the introductory section identify the critical problems and your main purpose?

Do you show all the facilities and resources needed for the proposed project?

Does the manuscript fill a need for the reader?

When one of the questions in the list reveals a weak spot in your manuscript, edit that section. Pare it down or build it up as needed. Then show the revised version to your colleagues for an independent check on your improvements in technical content.

After you have evaluated the content of your manuscript, review it for clarity, consistency, and accuracy.

Editing for Clarity and Integrity

Although you may not have the special skills of a technical editor, you can use the standard editorial approaches to ensure clarity of meaning

LIST OF CRITERIA USED BY JOURNAL REFEREES

Suitability

- Is the subject of interest?
- Is the work worthwhile, novel, unique?
- Has the work progressed sufficiently to justify publication?
- Would publication be timely?

Adequacy

- Is the reader properly oriented in the introductory section?
- Is the basic concept presented clearly?
- Does the author show adequate background knowledge?
- If data are given, is the volume sufficient and are the methods adequate to yield significant results?
- Are the vagaries of experimental conditions accounted for?
- Are the conclusions (or recommendations) based on the data given?

Balance

- Is the length of each section proportional to its importance?
- Is proper space devoted to interpretation and discussion?

Emphasis

- Are the significant results emphasized?
- Are the limitations of the results shown?

Previous Work

- Is appropriate credit given to others who have worked in the specific field?

and integrity of form and style. The writing defects selected here rank high on a technical editor's "hit list." Considering these will give you a start on the editing of your own work.

Begin by glancing through the document and comparing it with your topical outline. The main section headings in the manuscript tend to be similar to those in your original outline. If you did not prepare an outline before writing the manuscript, construct one now. It will be useful for reviewing the structure and especially for showing the sequence of your topics and their relative importance.

The basic structural requirement is that each section serve its intended purpose, as pointed out in Chapters 6 to 9. When you are editing, examine the functions of the separate sections. Does the abstract expose the core of the manuscript? Does the introduction give the proper background and indicate your purpose, etc.? When you are satisfied that the component parts do serve their intended purpose, look at how you have put them together.

In a well-written paper or report, the sections are interrelated by transitional devices. If necessary, add signposts that lead the reader smoothly from one subject to the next. Connective phrases, sentences, and paragraphs are the cement that unifies your piece of writing. If you insert them judiciously at the beginning or end of each section, the manuscript becomes far easier to read.

When you find a paragraph that does not read smoothly in logical sequence, look at the lead sentence, if indeed there is one. Consider the classical advice of the English instructor: Develop the main idea in each paragraph from its topic sentence. When you rethink and rephrase that sentence, you are taking the first step to clarify muddy exposition. You then have a basis for rewriting the paragraph.

On the other hand, the structure of only a single sentence in a paragraph may be the source of ambiguity. Identifying such a troublesome sentence in your own manuscript is half the battle. You may pass through the draft several times before detecting the twisted meaning in the sentence. When you recognize the defect, work on it. Rewrite for clarification—consider grammar, punctuation, diction, style, emphasis, etc. Rewrite the sentence several ways, considering every aspect. Editing and revising sentences may be frustrating at first, but eventually becomes the joy of effectively expressing yourself.

Finally, in reworking ambiguous portions of the text, do not overlook the favorite target of editors, *redundancy.* Lucid writing must be free of wordiness and repetition.

Up to this point your procedure has followed the practice of the professional editor, which is to review the structure and then to concentrate on successively smaller details: sections, paragraphs, sentences. If in your work situation you have no editorial services, you should proceed to ensure the integrity of the manuscript, reviewing it for consistency, accuracy, standardized forms, and stylistic requirements.

The first inconsistency you encounter may well be in the title, which must reflect any major revision of the text. Also, if you have a table of contents, be certain that it includes the final version of the section headings.

Whenever you revise the manuscript, you tend to create still other inconsistencies. As you add or delete figures, tables, equations, references, and footnotes, you disturb the numbering system. These changes must be checked carefully so that, for example, a new chart numbered "Figure 6" will match your explanation of Figure 6 in the text.

The captions for figures and tables should be specific, accurate, and self-explanatory. Compare the uninformative caption

Figure 4   Voltage-current curve for the autotransformer.

with the improved version

Figure 4   Relation between output voltage and digitized input current pulses
in the Xylon three-phase autotransformer.

When specific features of a graph or photograph are discussed in the text, it is sometimes helpful to add appropriate callout labels to the figure. Similarly, in tables the headings for rows and columns should correspond to the terminology in the text. Use of the same descriptive terms in text, figures, and tables thus aids readability.

Another matter of consistency is the accuracy of page numbers, which must correspond to the listing in the table of contents. Numerical values in the text should be double-checked for accuracy throughout, including those in mathematical equations and tables. Literature references need to be accurate, too. The complete bibliographic data should be compared with source documents for proper spelling and correct page numbers.

As in all engineering work, a professionally styled manuscript must adhere to accepted standards. The terminologies and technical abbreviations in certain specialties are standardized in publications of the engineering societies.

Examples of such reference works in the field of electrical engineering are

IEEE Dictionary of Electrical and Electronics Terms. IEEE Standard 100, 2nd ed., 1977.
IEEE Standard for Letter Symbols for Units of Measurement. IEEE Standard 260, 1978.

(These two standards are published by the Institute of Electrical and Electronics Engineers, New York.)

Similar publications in the field of mechanical engineering are

Abbreviations for Use on Drawings and in Text. Standard Y-1.1, 1972.
Glossary of Terms Concerning Letter Symbols. Standard Y-10.1, 1972.

(These two standards are published by the American Society of Mechanical Engineers, New York, and the American National Standards Institute, New York.)

In addition, engineering units are now well established in the metric system. For a guide to accepted usage, refer to a good compilation of the units.[52,63]

These various kinds of editing for integrity should be done on the draft. If your paper or report is to be typeset and you will receive galley proofs later, you should correct only the printer's errors on the proofs. At that stage in the publishing process, as pointed out in Chapter 23, editing for integrity will be a waste of time, money, and effort.

Chapter 20

How to Submit a Journal Manuscript Successfully

Selecting a Journal

When writing for periodicals, your objectives are to publish in the most suitable journal, to reach the widest interested audience, and to achieve both goals within a reasonable time. Unless you choose the right options, you will not reach those objectives.

A step in the right direction before you start to write is to develop a sensitivity to the needs of your audience and to consider a particular journal, as suggested in Chapter 2. When you are ready to submit a manuscript, it is wise to review once more your choice of a journal. Some better opportunities may have materialized since you started the paper. In addition, if you work for a company that has an editorial department, you can get some help in placing the manuscript.

There are important reasons that you should be flexible in your choice of a journal. The first is that several journals in your field may have almost the same readership but different policies for subject emphasis and structure of the articles, and you must find the one your paper fits best. The second is that some magazines, because of faster editorial processing, will publish your manuscript earlier than others. A journal that has scheduled a series of special issues on other topics probably will delay publication of your article for many months. Another factor in the choice of a journal is manuscript length. If your article is very short, you will have to select a journal that publishes "communications" or "notes." If your opus is very long, you must choose a publication that will devote the space to it. Otherwise, you will have to spend time and effort on requested revision.

Other reasons affecting your selection of a journal are the trends and fashions in publication. From time to time editors develop preferences for certain "hot" topics, to the exclusion of others. Your topic may now be at the top of a journal editor's list. Or it may be at the bottom because the editor just published several articles on your subject and now has no further interest in it. For these reasons you should scan current issues of your target periodicals.

One criterion for selecting a journal is the size of its readership. Do not overlook the opportunity for reaching a large audience. Small, specialized magazines can be ideal for reaching intended readers but may have a circulation of only two or three thousand. Others are widely read. For example, one engineering magazine, *IEEE Spectrum*, has a readership of well over 200,000. For a complete listing of circulation figures for periodicals, consult one of the standard reference works such as Ayer[7] or Ulrich.[12]

Another criterion is the professional reputation of the journal, judged, for example, by the frequency with which its papers are cited in other journals. A large number of reference citations is one indication of the "impact" of the journal in its field of engineering. Comparative listings for a large number of journals are published annually by the Institute for Scientific Information in *ISI Journal Citation Reports*. Figure 17 is part of a page giving a detailed breakdown of such citations for a one-year interval.

Cited references of your paper are, of course, only one type of evidence that your message is reaching its targeted audience. Reader response depends on the publication vehicle: the engineering society journal, the trade magazine, the conference proceedings, or the company journal. Before you decide where to send your manuscript, consider each of these four types— not only for the kinds of papers they publish, but also for the nature of reader response you can expect.

Engineering society journals. The periodical literature of the professional societies covers a broad range of engineering disciplines and is characterized by uniqueness and originality of concept. Topics include new theories and principles; development of devices, systems, processes, and materials; innovative designs and methods; measurement techniques; production methods; quality control; and contributions to the state of the art in other branches of engineering. Evidence of reader interest is found mainly in subsequent citations in the literature and in requests to the author for reprints. Usually each paper in an engineering journal is refereed, i.e., technically evaluated by a few authorities on the topic. A refereed journal offers two advantages: it provides the writer with expert opinion before publication and is a more prestigious vehicle for the paper than a non-refereed journal.

Technical trade magazines. The articles are generally devoted to new technologies, with emphasis on utility and practical applications. Topics include design methods, processes, and techniques of interest to engineers and management. Tutorial reviews, and also descriptions of the design and performance of new products, are sometimes included. Reader reac-

S C I   J O U R N A L   C I T A T I O N   R E P O R T S

CITED JOURNAL PACKAGE

	CITED JOURNAL / CITING JOURNAL	<-------- NUMBER OF TIMES THIS YEAR WAS CITED IN 1979 --------> TOTAL	1979	1978	1977	1976	1975	1974	1973	1972	1971	1970	REST
2.09	RADIOLOGY (CONTINUED)												
	6.50 ADV RADIAT BIOL	27	0	0	2	6	4	1	2	3	0	1	8
	.84 HEALTH PHYS	27	0	1	1	2	0	1	3	2	3	0	14
	1.69 INT J RADIAT BIOL	27	0	0	8	4	5	1	4	0	1	1	3
	3.65 AM REV RESPIR DIS	26	0	3	2	5	1	1	3	1	0	1	9
	.49 CAN J SURG	26	0	0	3	2	1	4	2	4	2	0	8
	.14 CATHETER CARDIO DIAG	26	0	1	2	0	1	1	3	1	1	0	16
	.88 DAN MED BULL	26	0	3	1	4	3	0	0	0	1	1	13
	.54 SEM HOP PARIS	26	0	0	3	0	1	1	3	3	2	1	12
	ALL OTHER (531)	3015	18	172	312	300	298	253	214	144	157	140	1007
.04	RADIOPROTECTION	6*	0	0	1	1	0	1	0	0	1	1	1
	.84 HEALTH PHYS	1	0	0	0	0	0	0	0	0	1	0	0
	.26 KERNENERGIE	1	0	0	0	0	0	1	0	0	0	0	0
	.77 MIKROCHIM ACTA	1	0	0	0	0	0	0	0	0	0	0	1
	1.09 NUCL INSTRUM METHODS	1	0	0	0	1	0	0	0	0	0	0	0
	NUCL TRACTS	1	0	0	1	0	0	0	0	0	0	0	0
	1.53 PHYSIOL BEHAV	1	0	0	0	0	0	0	0	0	0	1	0
.22	RADIOTEKH ELEKTRON*	906*	24	67	128	103	96	79	68	43	28	31	239
	.22 RADIOTEKH ELEKTRON*	617	18	43	97	70	68	57	49	31	18	21	145
	.37 IZV AN SSSR FIZ*	26	1	4	1	6	1	0	3	0	0	3	7
	.35 SOV PHYS SEMICOND*	21	0	2	2	0	5	0	2	1	0	1	8
	.50 KVANTOVAYA ELEKTRON*	18	0	1	5	2	0	1	0	1	2	0	6
	.31 ZH TEKH FIZ*	17	0	1	5	3	1	1	1	0	0	0	5
	1.03 ZH EKSP TEOR FIZ*	16	1	3	0	1	4	2	1	0	0	0	4
	.50 FIZ TVERD TELA*	13	1	1	0	2	3	2	0	1	0	1	2
	1.57 USP FIZ NAUK*	12	1	0	0	1	1	0	2	0	0	0	7
	.46 DOKL AKAD NAUK SSSR	11	0	2	1	0	1	3	0	3	0	0	1
	.04 TELECOMM RADIO ENG*	9	0	2	1	1	0	1	1	0	0	1	2
	.11 VESTN MOSK U FIZ AS*	9	0	0	1	2	0	2	0	1	0	0	3
	.42 FIZ ATMOS OKEANA*	8	1	0	1	0	2	0	0	0	0	0	4
	.32 GEOMAGN AERON*	8	0	0	2	3	0	0	0	0	1	0	2
	.24 IZV VUZ FIZ*	8	0	1	0	0	1	1	1	0	1	0	3
	.40 OPT SPEKTROSK*	8	0	1	2	0	0	1	0	0	0	0	4
	ALL OTHER (42)	105	1	6	10	12	9	8	8	5	6	4	36
.13	RAIRO-ANAL NUMER-NUM	6*	0	0	6	0	0	0	0	0	0	0	0
	.13 RAIRO-ANAL NUMER-NUM	2	0	0	2	0	0	0	0	0	0	0	0
	1.18 SIAM REV	2	0	0	2	0	0	0	0	0	0	0	0
	.43 NUMER MATH	1	0	0	1	0	0	0	0	0	0	0	0
	.76 SIAM J NUMER ANAL	1	0	0	1	0	0	0	0	0	0	0	0
.08	RAIRO-AUTOM-SYST ANA	3*	0	0	3	0	0	0	0	0	0	0	0
	.70 AUTOMATICA	1	0	0	1	0	0	0	0	0	0	0	0
	.73 INT J CONTROL	1	0	0	1	0	0	0	0	0	0	0	0
	NOUV AUTOMATISME	1	0	0	1	0	0	0	0	0	0	0	0
.16	RAIRO-INF THEOR-THEO	9*	1	2	6	0	0	0	0	0	0	0	0
	.16 RAIRO-INF THEOR-THEO	4	1	2	1	0	0	0	0	0	0	0	0
	.46 INFORM CONTR	2	0	0	2	0	0	0	0	0	0	0	0
	.34 ACTA INFORM	1	0	0	1	0	0	0	0	0	0	0	0
	.24 CR ACAD SCI A MATH	1	0	0	1	0	0	0	0	0	0	0	0
	.30 J COMPUT SYST SCI	1	0	0	1	0	0	0	0	0	0	0	0
.02	RAIRO-INF-COMPUT SCI	2*	1	0	1	0	0	0	0	0	0	0	0
	IEEE T PATTERN ANAL	1	0	0	1	0	0	0	0	0	0	0	0
	.02 RAIRO-INF-COMPUT SCI	1	1	0	0	0	0	0	0	0	0	0	0
.45	RCA REV	446*	4	3	21	25	58	31	25	34	22	40	183
	1.74 J APPL PHYS	36	0	0	0	1	10	2	3	2	2	2	14
	1.87 J ELECTROCHEM SOC	29	0	0	0	3	2	0	1	0	1	10	12
	1.67 IEEE T ELECTRON DEV	22	0	0	1	4	3	1	3	2	1	1	6
	.45 RCA REV	19	3	0	5	3	3	0	2	1	0	0	2
	1.53 TOP APPL PHYS	18	0	0	1	1	4	3	0	0	0	1	8
	3.26 APPL PHYS LETT	17	1	0	2	0	3	1	1	0	2	4	3
	1.08 THIN SOLID FILMS	13	0	0	1	0	1	0	0	0	0	4	7
	1.02 IEEE T MICROW THEORY	11	0	0	1	1	1	0	1	2	0	2	3
	.76 JPN J APPL PHYS	11	0	0	0	1	4	0	0	0	2	2	2
	1.34 SOLID-STATE ELECTRON	11	0	0	0	0	2	1	0	0	0	1	7
	1.57 APPL OPTICS	9	0	3	0	1	0	0	0	3	1	0	1
	.60 ADV ELECTRON EL PHYS	7	0	0	0	0	0	0	2	0	0	1	4
	.50 FIZ TVERD TELA*	7	0	0	0	0	2	0	0	3	2	0	0
	1.79 IEEE J SOLID-ST CIRC	7	0	0	1	0	0	0	1	1	1	0	3
	.93 SOLID STATE TECHNOL	7	0	0	2	0	0	0	0	1	0	1	3
	.35 SOV PHYS SEMICOND*	7	0	0	0	0	0	0	0	0	0	1	6
	.97 IBM J RES DEV	6	0	0	0	0	1	0	1	0	0	0	4
	1.42 J ELECTRON MATER	6	0	0	0	0	4	0	0	0	0	1	1
	1.01 MOL CRYST LIQUID CRY	6	0	0	0	0	0	4	2	0	0	0	0
	.51 REV PHYS APPL	6	0	0	2	1	0	0	0	0	0	0	3
	3.92 SURFACE SCI	6	0	0	1	0	1	0	0	0	1	0	3
	ALL OTHER (103)	185	0	0	5	8	17	19	8	19	9	9	91

Figure 17 Specimen listings from *ISI Journal Citation Reports.*

tions can be gauged by the number of requests to the author for information and by the demand for reprints. A strong reader interest is indicated if the editor features the article on the cover. If the editor highlights your article in a special section, you can also expect it to attract attention. The advantages of choosing a trade magazine are relatively quick publication, wide circulation, and payment for the article.

Conference proceedings. Any large engineering conference, symposium, or convention usually publishes the papers presented, either in a condensed version or in full text. There is a considerable range of topics, depending on the kind of meeting:

- *Engineering society conferences or symposia.* Proceedings are similar in content and technical level to society journals. Some papers, however, may offer only preliminary findings and thus will be less comprehensive than a journal paper. Proceedings also tend to include more review papers than do the journals.
- *Trade conventions.* An engineering convention may include technical sessions. The papers feature new technologies and the properties of new products and systems. The technical level varies according to the requirements of the program organizer.

Presenting a paper at a conference has advantages over publishing in a journal. First of all, it allows direct interaction with the audience. You usually receive feedback from your peers during the discussion period. You also have the opportunity to make copies of your text available at the meeting. Questions from the audience and the number of reprints picked up after your talk are good indications of interest. Later on, evidence of reactions to your published proceedings paper will be similar to that for a journal paper.

Company journals. Some large engineering companies publish technical journals reporting work in their laboratories. Many such journals are product oriented and publish descriptive articles. Others, such as the *Bell System Technical Journal* and the *IBM Journal of Research and Development,* publish primary R&D papers. Reader response differs from that to society journals and trade magazines. For example, the internal circulation of the journal stimulates more inquiries from the community of engineers within the company. These inquiries and the resulting discussion and cross-fertilization of ideas are evidence of reader reaction. External circulation also results in citations in other journals, as shown in the listings for the *RCA Review* in Figure 17.

Manuscript Requirements

To succeed in placing your paper, you must do more than send a neatly typed manuscript and a letter of transmittal. You will have a better chance of acceptance if you anticipate possible reasons for rejection and if you understand the needs of editors. Typical engineering journals may reject 25 to 40 percent of submitted manuscripts. Of those papers that are accepted, some encounter extensive delays before publication as a consequence of refereeing, slow editorial processing, or the complexities of production scheduling. If the manuscript is of immediate interest or reports a significant new engineering development, these delays are frustrating. You can sometimes avoid such difficulties if you take certain precautions when submitting the manuscript. I suggest different procedures for the four types of journals because each has distinctive requirements.

Engineering society journals.   The journals of the engineering societies have fairly straightforward instructions to authors. These are usually printed in each issue as "Information for Contributors," or the editor will supply a pamphlet. Typical information includes the format in which the manuscript should be typed, the style of illustrations, the number of copies required, and the mailing instructions. The criteria for accepting a manuscript may or may not be included. You may wish to telephone the editor and ask about (1) the average lead time to publication, (2) any special issues that would have priority and thus delay your paper, and, unless already specified, (3) the editor's criteria for acceptance. If any of these three items seem to present a problem, try the next choice on your list of preferred journals.

Technical trade magazines.   For a trade magazine it is best to get in touch with the editor before you send the article or even before you write it. Get agreement from the editor on the subject and your treatment of it. Usually a telephone call is sufficient. If not, the editor will ask you to send preliminary material: title, abstract or outline, estimate of figures and tables, and an indication of why it will interest the readership. If the editor then shows interest and suggests changes, you have almost sold the manuscript before it is written. You can then suggest material for a cover design to feature your work, if you think it appropriate. Ask about deadline requirements. You have then established a good relationship with the editor, not only for the present article but also for future ones you may wish to submit.

Conference proceedings.   Proceedings have different requirements. The call for papers is not always specific and may give only the general subject areas. Get in touch with the program organizer and inquire about the criteria for acceptance, the kind of emphasis desired, and the limitations of scope. Then send in the abstract of your paper before the deadline, with close attention to those criteria. Your contribution to the program may be accepted on the basis of the abstract, or you may be asked to submit the full manuscript for review. After acceptance you will be asked either to send your final verison for typesetting and printing or to prepare that version on reproduction form sheets, which will then be sent to the publisher. Because these forms will be made directly into offset printing plates, you assume the sole responsibility for letter-perfect copy. If the proceedings is to be available at the conference, the printing schedule will be tight. Send your final manuscript promptly.

Company journals.   The circulation of company technical journals is usually both internal and external. Publishing procedures differ from those of outside publications. The editorial policies are guided by company policies, which can be an advantage to an author reporting on new development work. For example, company confidential technical reports are sometimes declassified and eventually submitted to outside journals. But if they are submitted to the company journal while still classified, the editor can review them and prepare them for publication parallel to the declassification process. Thus when they are finally cleared, they can appear immediately in the company journal instead of waiting several more months for review and publication in an outside journal. If you work for a company that publishes a technical journal, discuss the timing of your article with the editor. One other advantage of the company journal is the extent to which the editor is willing to publish a number of the company's engineering papers on the same topic in a prestigious special issue.

Generalized Procedures

If your selected journal does not have special instructions for format, the following procedures will lead to a generally acceptable manuscript.

Examine the current issue of the journal to determine its standard format. You may need to supply the following items: author's address, an abstract, and a biographical sketch and photograph. Be sure to follow the journal's scheme for numbering and capitalization of section headings and subheadings, the form of literature references and footnotes, and the method of citation of references in the text.

Your manuscript should be typed double-spaced with generous margins on one side of the paper only. Each page, including figures and tables, should be numbered. A list of all figure captions belongs on a separate sheet of paper.

Important mathematical equations, either typed or legibly written in black ink, should be placed on separate lines in the text and numbered consecutively. Take extra care to place superscripts, subscripts, and operation symbols properly. Place a wavy line under boldface characters such as vector and tensor quantities.

When photographs are included, use glossy finish. Pictures must be sharp and of good contrast. If labels and scales are to be added, indicate them clearly on a separate sketch or photocopy. Identify each photograph on the back, and, if necessary, indicate which edge is the top.

If the journal does not redo line drawings, provide clean, even linework and accurately worded axes and labels.

Proofread carefully, following the suggestions in Chapters 19 and 23 for ensuring the integrity of the manuscript.

Steps to Publication

Prepare a letter of transmittal addressed to the journal editor. Your letter should simply state that you are submitting the manuscript for consideration. Enclose three unfolded copies of the manuscript, and send them by first class mail. If you do not receive an acknowledgment within two weeks, follow up with a phone call or a second letter.

The subsequent procedures for reviewing, editing, and publishing an engineering manuscript depend to some extent on the kind of journal. In general, after the refereeing of the manuscript, the editor returns it to the author with either the verbatim reports of the reviewers or a summary of their comments. It is the editor, and not the reviewers, who ultimately accepts or rejects the paper. Subsequently, the editor makes the final judgments on the adequacy of an author's alterations to the manuscript. After acceptance the paper is copy edited by the journal staff, and galley proofs (or in some cases page proofs) are sent to the author for final proofreading. Corrected galley proofs are set into page format for final proofing by the editor and for printing by the production department of the vendor.

A trade journal editor will usually evaluate the article but may consult with other staff editors or, on occasion, with an outside referee. Trade journal editors tend to do more editing than society journal editors and frequently rewrite portions of the manuscript.

Publication in a conference or symposium proceedings involves prior evaluation by the program committee. The judging procedures of various program committees vary considerably. Some accept a paper solely on the basis of a submitted abstract. Others, like editors of professional journals, resort to extensive refereeing.

One procedure is common to all four kinds of publications. If, after technical evaluation, the editor (or program committee) suggests changes, the author can either make the requested revisions, explain why the suggested alteration is not valid, or withdraw the manuscript.

A referee's report demands careful attention and interpretation, and the best course for you to pursue is the subject of Chapter 21.

Chapter 21

How to Deal with Your Critics

The Review Process

The techniques described in this book are designed to help you write the kind of report or paper that will pass with flying colors the test of peer review. If the work reported is sound, any reviewer who appraises your manuscript for technical content and writing effectiveness will react favorably to certain hallmarks of quality. For example, reviewers appreciate the value of an abstract that immediately lays bare the heart of the paper, i.e., the nature of the problem, the approach taken, and the essential results. They also quickly sense the usefulness of an introduction that reveals the motivations for the engineering project and its relation to the existing literature. Any reviewer will feel comfortable with a paper that emphasizes and expands upon its more important contributions while devoting less space to peripheral explanations. Experienced evaluators know how an audience is attracted to a paper that shows empathy for the needs and interests of readers. The other elements of quality described in these chapters will also help to create an aura of acceptability.

Peer review, however, can be a complex process. For example, if you decide to submit your manuscript to a refereed journal, the editor will usually appeal to two or three referees for independent opinions. As pointed out in Chapter 20, the editor's acceptance is based on the critical judgments of these experts, who are not on the editorial staff. Such technical reviewers may serve in the editor's corps of referees or they may perform a one-time service. The editor is the middleman who communicates first with the referees and then with the author.

The identities of referees are usually withheld from the author, which removes the personal element from refereeing. Some editors, feeling that authors should also be anonymous, remove the authors' names and affiliations from manuscripts.[19] Most, however, avoid this practice of "blind review," as explained by Juhasz et al.[30]:

> ... editors are aware of how well referees can frequently discover who the author is from his theoretical approach or nature of the experiment. Also, a reviewer might be offended if he received a paper this way and felt that the editor suspected him of bias and had not provided full working information.

The customary anonymity, then, is one-sided. Although, for the most part, referees are fair and objective, they can conceivably be influenced by knowing who you are. But even if the referees are guilty of bias when voting against your paper, there is a justification for their remaining anonymous: the secret ballot is a feature of the democratic process.

You will be better able to analyze the evaluations and defend your paper if you understand the choices an editor must make and the vagaries that plague the editorial process. The responsibilities of the editor to choose material that will uphold the quality of the journal are twofold: (1) to assess manuscripts for technical and literary merits and (2) to confine the selection to topics within the range of reader interests. In principle, referees also are sensitive to both these needs, and their opinions serve as a test run of reader reaction. But on occasion the editor will be a better judge of the readership and will decide on the merits of a manuscript accordingly.

Journal editors differ in the kind of guidance they give referees. Some give minimal instructions, trusting to experience and mature judgment. Other editors use rating forms of the kind I used for many years (see p. 124).

Referees may be asked to prepare commentary in a form the editor can send verbatim to the author. A reviewer's report written for this purpose will usually lack harsh language and hypercritical judgments. Or the referee may be required to give only an informal and private critique which the editor then summarizes and interprets for the author. An editor's decision to send you only a summary generally has one of two purposes: to protect your sensibilities against the barbs of critical referees or to clarify their reports by deleting irrelevant minutiae.

The Referee's Role

A referee's authoritative appraisal serves to maintain the quality of the journal and also to protect your professional reputation. A highly critical referee who finds weaknesses, errors, or improprieties is thus acting as your best friend. Good advice gives you the opportunity to improve a basically good paper or to withdraw a basically bad one. If you are overly defensive, however, you may misinterpret the referee's report.

It is helpful to understand the role played by a referee and the viewpoints that affect the critique offered by an expert. The viewpoints of a journal referee are similar to those of reviewers of conference papers, internal company or government reports, or students' theses.

The first viewpoint arises from the expertise of the reviewer, whose own background of experience in the topic is important. If the subject is a narrow specialty, this point may be especially significant. An expert in one field of engineering is not necessarily qualified to judge a paper in a related discipline. A negative report from such a referee may or may not be valid.

A second point of view is the reviewer's attitude about the existing literature on the subject. A reviewer either should know the papers and reports published on the topic or should be willing to research them before writing a referee's report. The reviewer should indicate whether or not significant references have been cited by the author.

One of the most important points of view is the referee's perspective on the validity and adequacy of the work reported. Some referees base their judgments on the importance to other engineers doing similar work. Others appraise only the value of the paper to the general reader of the journal. Still others attempt to strike a balance between these two criteria. When referees have differing views on this point, they can disagree sharply on the merits of the manuscript.

Another perspective of referees concerns the literary qualities of the paper. The author's ability to communicate is, of course, essential to any manuscript, but some reviewers are guided by personal preferences, rather than correctness, when criticizing diction, punctuation, and even the general organization of a paper.

The remaining viewpoint concerns the attitude toward the author's work and the subtleties of professional ethics. A referee should be courteous and considerate of an author's efforts no matter how poorly written the manuscript or how inadequate the content. In addition, an honest appraisal of the paper should be given for the benefit of the journal and the author. A referee must avoid unfair bias in judging a manuscript that is in competition with his own engineering development. From an ethical point of view, it is just as important to avoid favoritism toward the author when the work reported merely supports the referee's own pet hypotheses.

These remarks about the viewpoints of a referee suggest that you need not always accept negative criticism at face value. The underlying reasons for the weak arguments against your paper may lie in the attitude of a referee. A careful analysis of the report can then support your reply to the editor.

The Author's Role

Evaluating the Evaluator

Although the referees' evaluations are usually helpful, you should review all of their comments with great care and take advantage of fresh viewpoints on your writing and your work. Remember that negative comments can be highly constructive when they reveal the weaknesses in your manuscript and indicate how you can strengthen your final version.

If, however, the referee's suggested changes seem unreasonable, it is your turn to evaluate the evaluators. There are several reasons why you should attempt to assess your critics:

- As an engineer you are equipped with analytic skills. Probe the evaluation sent to you, which may be either the verbatim reports of the referees or the editor's summary. Look for tell-tale clues of misunderstanding, bias, trivial "show-off" remarks, and even incompetence. But be objective. It is too easy to reject valid comments about your brainchild, the manuscript.
- Although independent evaluations are a tool for editorial decisions, the chosen referees may not be the ultimate authorities on your topic. Indeed, the selection of *ideal* referees for every manuscript is extraordinarily difficult for an editorial staff that processes hundreds of manuscripts per year. Besides, referees and editors are fallible.
- The editor will be interested in your defense and rebuttal. If your reasoning is sound and the facts correct, the editor will admire your integrity and besides can use your rebuttal as a rationale for reversing a negative decision.
- Your own analysis, probing into those first two sets of referee-editor opinions, *can actually be the most revealing of all* when combined with your own insights about the manuscript.
- The editor may decide to have you get in touch with the reviewer directly, with the reviewer's consent. A polite exchange of views on the merits of your manuscript will usually be an educational and clarifying experience for both you and the referee. Polemics, however, will seldom accomplish anything.
- Even if the editor does not insist on revisions, you must assume the responsibility for clarifying your writing. If you ignore the suggestions entirely and your paper should appear in the journal unrevised, some readers could have the same reactions as the referees to the portions in question.
- Resolving questions about your paper can save you many months in getting it into print, because sending it to a different journal requires more refereeing and editorial processing.

The Editor's Summary

If the editor sends only a summary of the referee's report, your first approach should be to appraise the tone of the editor's letter. If it is written in a sympathetic and open-minded vein, with suggestions for remedying the deficiencies, you can assume that your response will be given every consideration. However, if the letter hints that your paper does not seem suitable for the journal, or if it cites only vague technical or stylistic reasons for the rejection, you had best submit your paper to another journal immediately.

Instead, the letter may encourage you to revise. If you feel the suggested changes are unwarranted, examine the summary and try to separate the editor's opinions from those of the referees. An editor's own criticisms must be answered with tact. Never contradict those judgments or argue against them. Instead, the best way to reverse an editor's opinion is to show (perhaps indirectly, but unmistakably) that additions or changes to your paper will stimulate the interest of readers. For example, an editor might raise the following objection:

> The design of your dual mirror system is too complex for practical use.

To reply you can insert in the manuscript two sentences that acknowledge the complexity, but also identify the needs of readers and cite low-cost practicality:

> The main design problem in solar systems today is the low efficiency of collecting solar radiation. The complex configuration of dual mirrors in this collector gives a 35% improvement in thermal efficiency over conventional types and can be mass produced with a single low-cost mold.

The technical objections of a referee, as cited by the editor, can frequently be refuted by adding explanatory information. In this example, the referee attacks the validity of your results:

> There is no known physical reason for the corrosion rate to be related to the ratio of polyethylene to water.

In your response you might add the following statements, indicating why the objection is irrelevant:

> Although the effects of the ratio of solutes on corrosion are not widely recognized in the literature, the data in Table 3 show how the ratio can be an important factor. These data strongly support the tentative conclusions reached by Jones in Ref. 6.

Thus, if the editor's summary asks for specific clarifications, you have the opportunity to upgrade your manuscript. But the unedited reports of the referees can also be illuminating.

The Referees' Reports

When the editor decides to send you the full reports of two or more referees, you can make a deeper analysis than on a summary. If you are to make an effective response to adverse criticism, you must study each referee's report to determine its main thrust, which usually concerns technical merit, suitability for readers, or writing skills. Give special attention to the referee's technical questions. Go over them with a colleague so that you can develop a balanced view.

After you have considered the comments and questions in the light of the referee's probable motivations and interests, you will be better able to respond to the editor. As you are examining the report and probing its meaning, use the following questions as a checklist for your analysis:

- *Are the comments plodding and mechanically stated or are they imaginative and challenging?* Give more weight to the referee's remarks that indicate a special interest in your topic. The generalized comment, which could apply to almost any engineering paper, has the least value. A complaint that your work is "poorly organized," without a suggestion for reorganizing the structure, is a lazy and useless criticism.
- *Are all of the referee's points negative?* A reviewer who does not acknowledge the beauty of the best parts of your paper is probably prejudiced and is merely attempting to impress the editor with his powers of critique. You should look for some compliment, however veiled, on your manuscript. If all comments are negative, your referee is making no effort to be helpful (assuming, of course, that your manuscript does have merit).
- *Does the referee cite facts when criticizing your concepts, methods, or results?* You need not give credence to speculations, unsupported hunches, or professional snobbery.
- *Are the remarks mostly trivia?* The most valuable observations are the referee's assessments of the validity, current significance, and general quality of the manuscript. A few authoritatively stated remarks of this kind will outweigh a large number of carping comments.
- *Are reasons given for the details of the overall appraisal?* If the paper is excellent, the referee should state what it contributes and why it should be published. If it is poor, the review should point out the defects and errors. Any such criticism should be constructive if it is to be useful to

the editor and author. The absence of the reasons is good cause for your queries to the referee.

- *Does the referee contribute new insights to the paper?* Some of the best reviews aid the author by offering additional interpretation of the findings.
- *Does the referee expect answers to the questions or revisions to the manuscript?* A professional opinion is not intended to be criticism for its own sake. Look for an illumination of the ideas in the manuscript and an opportunity to exchange thoughts and to ensure the quality of the final version.

Your Alternatives

Do not be carried away by the glowing approvals of referees. Search out their helpful suggestions for further improvement.

If a referee's reports are negative, the editor will either reject the manuscript or request that you alter it. Here is a review of your choices:

- *Make the requested changes.* It is unnecessary, however, to follow all the minute suggestions of referees. Be sure to assess the points of agreement and disagreement in the independent reviews. For example, among two or more referees agreement on relevant literature will probably be high and agreement on reader interest will probably be nil, as noted by Williams.[65]
- *Defend a rejected paper.* After carefully analyzing the reports, you may still feel that your paper is defensible. Write a polite rebuttal to the editor. Present a strong case supported by facts and revealing (only by implication) the distortions that may have crept into referees' comments as a result of misunderstanding, bias, poor judgment, or sheer arbitrariness. All editors admire and respect a well-reasoned rebuttal. In rare cases, some journals even permit author involvement in open peer commentary, as pointed out by Peters and Ceci, in which "creative disagreement is published in its entirety for all readers of the journal to see and appraise."[51]
- *Inquire about the reasons of your critics.* The editor will often arrange for you to discuss the matter directly with a referee.
- *Submit the manuscript elsewhere.* But only as a last resort.

Chapter 22

How to Review
Engineering Manuscripts

It is important to understand that the development of your writing skills and the plans to have your various results published will continue after your first journal paper appears. In unexpected ways, as you mature as an experienced author your professional activities will enhance those skills and actually help you to publish successfully.

An example of one of these activities is your role in appraising manuscripts written by colleagues. In today's complex world of technical developments, engineers seldom work alone. As part of a working group or an operating department you will occasionally be asked to look over the draft of a report or paper written by one of your co-workers.

After you have published a few papers in journals, you will become better known as a contributor in your field—perhaps even an authority— and a journal editor may ask you to review a manuscript recently submitted.

At first you may agree to act as a referee out of a sense of duty and will feel that appraising manuscripts is merely tedious and time consuming. Gradually, however, you will realize that serving as a critic can help your own writing and publishing efforts. Here are some of the peculiar benefits you may not have anticipated:

- Writing the kind of referee's report described in Chapter 21 will sharpen your critical faculty. This sensitizes you, more than ever, to structural faults and technical weaknesses in manuscripts. In subtle ways, these increased sensitivities will strengthen your future writing and your ability to criticize your own work.
- Reading novel and unpublished results in your field becomes a fruitful source for developing new ideas and interpretations in future papers of your own.

- Becoming a referee and offering advice establishes a new relation between you and the journal editor. This tends to give the editor more confidence in papers you submit in the future.

As a referee you will be giving a technical assessment for the editor and also providing a typical reader's reaction to the paper. Moreover, in judging the value of the paper and its suitability for publication, you help to protect the integrity and reputation of the journal. At the same time you will be rendering a service to the journal's readers by screening material according to their needs and interests. Finally, your review may aid the author by giving constructive suggestions for upgrading the manuscript and perhaps the engineering work itself. And so your recommendations can be an indirect contribution to the engineering community.

The purpose of this chapter is to show specifically how to review an engineering manuscript and also to show that the character of your evaluation will depend on your sensitivity to the author's objectives, the validity of the work reported, and the journal audience.

Your Role as a Critic

The very first step in your review procedure should be to look at the manuscript and decide whether you are the right person to review it. If you are not, return it to the editor promptly. If you do feel qualified, the next important step is to scan a few recent issues of the journal (unless you read it regularly). The current selection of papers published is evidence of the present editorial policies, which may have changed recently. The table of contents also indicates the interests of the readership.

If you do not fully understand these two items—editorial policy and reader interests—you will be poorly equipped to review the manuscript. Of course, the editor had made an initial assessment on suitability before sending it to you. The editor, however, always depends on you for a "fine tuning" of that preliminary judgment. To attempt to review the paper on its own merits without examining the journal, then, is a thoughtless way to proceed, regardless of your judging criteria.

Whether or not you have wide experience as a peer reviewer (referred to by most editors as a "referee"), you do need a set of criteria for evaluating manuscripts. When a paper is sent to you for review, a standard rating sheet or a general instruction is usually included. However, some editors will offer no guidance, blandly assuming that you know all the essentials of the art of refereeing.

Even if the editor should specify criteria for your review, I suggest that you examine the list on page 124. As a journal editor, I have been sending this set of questions to referees for over 25 years and find it useful.

The list is not intended as a rating form but rather as a general guide for writing a formal critique. I do not favor numerical rankings or yes/no judgments of criteria. There is simply no substitute for opinions expressed in your own language and fueled by your background of experience. If an editor does send you a multiple-choice rating form, you should comply but also add a running commentary.

The list on page 124 may not offer you new ideas on how to assess a manuscript for style and content, but at least it serves as a memory aid, reminding you what to look for in your search for strengths and weaknesses.

The Criteria

Before examining the details of the paper you should first decide about its general suitability for the readership. The subject, of course, must fall within the range of topics ordinarily covered in the journal, and the technical level of the writing must obviously be appropriate. But there are other aspects of suitability. For example, would the paper fill a need for readers and is the author's new design technique actually of current interest? Remember that accuracy and validity of a paper are not sufficient reasons to publish in a particular journal if the information seems of no use to readers and would have little value to them for future reference.

If the paper reports new information of some kind—either descriptive or analytic—the author should refer meticulously to previously published results. The way a writer mentions prior articles always deserves close examination. For example, are the deficiencies and limitations of previous work shown? Are important contributions to the literature acknowledged? Are the references sensibly selected and accurately cited? Is their relation to the author's present work clearly indicated? The manuscript should satisfy these questions. The author of a weak paper (at least in some kinds of engineering magazines) often ignores or even evades the issue of prior work.

Another question concerns the manuscript as a whole: Does it adequately serve its purpose? Engineering manuscripts may describe methods for designing a device or system, show properties of engineering materials, offer new measurement techniques, analyze problems, make recommendations for new projects, etc. Whatever the purpose, the paper must clearly define its basic concept, offer accurate data, and give valid conclusions. An assessment of the way these three elements of information are organized in the manuscript is one of your main responsibilities.

One way to determine whether the manuscript is logically organized and whether its important points are properly emphasized (see Chapter 10) is to examine the relative length of the sections and the author's choice of figures. When too much space is devoted to peripheral explanations and too many side issues are illustrated in the figures, you will sense that the paper is out of kilter and you should recommend ways to bring it into balance.

One of the most important criteria is the way the data are presented. The statistical treatment of the data usually deserves careful inspection[51] because in many cases it is the main support for the author's contribution.

Your Role as a Contributor

Your report to the editor is, of course, a personal opinion and the commentary need not be confined to the items listed on page 124. Sound judgments will be partly based on your intuition about the actual value of the manuscript. Constructive suggestions can be a real help to the editor and can become an anonymous contribution to the paper when you cite items like the following and suggest how the author can make changes or additions:

A significant technical omission in the manuscript.
An important error, experimental or analytic, in the engineering work.
A new and different interpretation of the author's results.

When equipped with your observations the author has the opportunity to upgrade the paper. Revision may require that an experiment be redone to get better data or more information. Or your report may influence the author to do some rethinking and rewriting that will better illuminate the paper with the ingenuity and beauty of the engineering accomplishment. In any case publication of the revised paper will give you the satisfaction of having contributed useful ideas.

A journal editor will quickly appreciate the difference between such a referee's report and one that only provides numerous trivial comments. The editor is the final judge, who compares and weighs the reports of the referees and makes a final decision, however complex,[26] on the pros and cons of accepting the manuscript.

The Value of Your Report

One way to summarize my advice is to point out some of the qualities of good and bad reviewing. Indeed, a referee's report can be appraised by the journal editor and the author in somewhat the same way as any technical manuscript, as pointed out by Gordon.[25] Compare the characteristics of good and poor reviews:

Good Reviews

Definitive analysis
For the journal editor a careful evaluation by an expert referee will define the virtues and defects of a submitted paper.

Creative suggestions
A referee's critical comments, either positive or negative, can be a source of technical improvement, and even of new ideas, for the author to inject into the manuscript. Thus a critique, in addition to identifying technical errors and stylistic defects, can provide fresh insights.

Intellectual stimulus
Evaluating papers can be a chore but may also turn out to be a boon for the referee. Having an early look at unpublished work offers the critic advance information on new developments. Moreover the expert who is called upon for a professional opinion can find an intellectual stimulus in new concepts and novel refinements of previously reported work. Of course, a referee is expected to adhere to the ethics of reviewing by respecting the author's rights of confidentiality prior to actual publication of the paper.

Poor Reviews

Excessive fault finding
An intensive search for superficial faults can be used deliberately as a basis for rejecting a manuscript. But when indulging in such trivia the referee may overlook the chief technical virtues of a paper.

Prejudice or bias
The referee's preconceived notions or preferences can color the critique. The professional practice, of course, is to review papers on a high ethical plane,[54] but subconscious bias can creep into the evaluation.

Unfair generalities
A referee's recommendations, when phrased in general terms and unsupported by statements of the reasons, are not helpful. This is true for either favorable or unfavorable comments like "This paper is a contribution and should be published" or "The author's conclusions are not valid." Such remarks will be of little value unless the referee explains what the author's contribution is or why certain claims are not valid. Even though editor and author should accept the judgments offered by an expert, the *rationales* for those opinions will be illuminating, especially when the independent judgments of two referees are poles apart.[15,27,65]

However, the ultimate consideration for the technical review is its value to the engineering community, i.e., the journal readers. And a referee's report also serves an excellent purpose when it offers constructive help to enhance the professional reputation of the engineer author.

Chapter 23

How to Proofread

The Author as Proofreader

It is important for you to know how to proofread the final form of your report or paper. If it is a thesis or engineering report, the final version may be either a typescript or the printed output of a word-processing system. If it is a journal paper, it is usually set into type, and the printer's proofs you receive will be either single-column galleys or page forms.

A typist, word-processing operator, editor, or printshop proofreader will be checking the finished version (typed, printout, or typeset) for errors. Usually you also will have the opportunity to do so. Or you may be the *only* proofreader if your typed manuscript is to be reproduced by offset printing in a conference proceedings or a company report. When you have co-authors, one of you will generally assume the responsibility for proof-reading.

Authors are sometimes poor proofreaders because they are so close to the work. However, an author has one distinct advantage over others who check for inaccuracies and mistakes. The author's eye is more sensitive to certain kinds of errors and defects that elude proofreaders who check only against a previous version of the manuscript. And although a technical editor is trained to find inaccuracies, only an author can recognize the type of factual error that may go unnoticed by others. Authors, for example, are expected to correct in the first draft the misplaced decimal points and the numbers miscopied from an engineering notebook or other source of data. An author who did not detect such blunders in the manuscript may recognize them in the printer's proofs, which always provide a writer with a fresh perspective.

There are, however, other reasons why you should read the final versions of the manuscript and the printer's proofs. You should proofread a copy editor's last-minute alterations. And if you have a colleague who is an especially good proofreader, enlist his or her help for a separate reading. For journal papers, meticulous proofing will save both author and publisher from the awkwardness and embarrassment of having to print errata in subsequent issues.

Reading Manuscript and Proof

Authors tend to confuse proofreading with self-editing. The confusion causes extra work, delays, and unnecessary expense. Some definitions clarify the distinction:

- *Editing* involves deleting, altering, or adding material to the early versions to improve their literary and technical qualities.
- *Proofreading* is quite different; it involves searching for errors in the edited version. Your purpose is to provide the ultimate readers with a letter-perfect report or paper, free from the mistakes that occur in typing, type composition, or preparation of artwork.

You must do your own editing before your manuscript reaches its final form. If you delay the self-editing until you receive galley proofs from a journal's printshop, the alterations become costly. Besides, too many "AAs" (author's alterations) in the proofs require changes in type composition that tend to propagate still more errors in the composing room. The result is additional generations of proofs and more corrections. Most publishers will bill you for AAs.

The Camera-Ready Manuscript

If the final form of your report or paper is typed or word processed instead of being composed in type, it can be used as a camera-ready document to be photographed, processed, and then printed in quantity by photo-offset. No matter how carefully a typist prepares the camera copy, you should also proofread it yourself. After several readings of the various stages of the draft, however, you become too familiar with the text. You will tend to read from memory and miss some of the flaws that find their way into the camera-ready copy. To avoid "memory reading" wait at least a day or two so you can make a fresh approach to proofreading.

The best method is to have someone read the draft aloud slowly as you check the camera-ready copy sentence by sentence. If instead you decide to proofread alone, go through the manuscript at a leisurely pace at a quiet location where you can concentrate on it, away from telephones and other distractions. When you cannot avoid interruptions, lay a ruler under each line in the draft and another on the camera-ready copy as you compare them. This will help you keep your place.

Have a dictionary at hand to check spelling and hyphenation. Take special care with technical words and abbreviations that may not be familiar to the typist. Misspelled and misused words are not the only errors that plague proofreaders. Even the best typist can occasionally skip a line or neglect to start a new paragraph.

Your previous editing in various stages of the draft, as suggested in Chapter 19, should have weeded out all inconsistencies, but any subsequent changes require attention to the following in the camera-ready copy:

- The consecutive sets of numbers for figures, tables, references, equations, and footnotes.
- Reference in the text to figures, appendices, etc.
- The data and line work in illustrations.
- Phrasing of headings and captions in tables and figures.

The Printer's Proof

Although some journals reproduce camera-ready manuscripts directly, most use some form of typesetting (usually photocomposition), and the editor will send you either galleys with separate proofs of the figures or else page proofs, i.e., the final layouts including figures and tables.

The printer's proof is the product of a two-step process: an editor's preparation of the manuscript and the compositor's typesetting. Such a two-step procedure is subject to human error, even with skilled copy editors and type compositors.

Your procedure is similar to reading camera-ready copy, but you need to take several additional precautions:

- Make all corrections in the margins of proofs as shown in Figure 18.
- Check the headings and captions once more. Although editors are adept at revising them, you should review the changes for accuracy.
- Give special attention to typeset equations. Symbols like integration $\int$, summation Σ, and partial differentiation ∂ should be of appropriate size, properly located, and sensibly spaced. When the display equation is too long for a single line, the break should occur preferably just before a verb ($=$, $\leq$, $\geq$) or else just before a conjunction ($+$, $-$). Be sure that superscripts^a, subscripts$_b$, and their modifying indices$_{b_y}^{a^x}$ are positioned correctly. Built-up fractions $\dfrac{1}{x + y}$ should be avoided in text and replaced by a solidus fraction $1/(x + y)$. Remember that you will be more sensitive to errors in the translation of mathematics from manu-

PROOF CHANGES

Marks	Explanation	Marks	Errors Marked
ℰ	Take out letter, letters, or words indicated.	ℰ	He marked the proof.
#	Insert space where indicated.	#	He marked theproof.
ℰ	Turn inverted letter indicated.	ℰ	He marked the proof.
ℓ	Insert letter as indicated.	ℓ	He maked the proof.
lc	Set in lower-case type.	lc	He Marked the proof.
wf	Wrong font.	wf	He marked the proof.
×	Broken letter. Must be replaced.	×	He marked the proof.
ital	Reset in italic type the matter indicated.	ital	He marked the proof.
rom	Reset in roman (regular) type the matter indicated.	rom	He marked *the* proof.
bf	Reset in bold-face type word, or words, indicated.	bf	He marked the proof.
⊙	Insert period where indicated.	⊙	He marked the proof
tr	Transpose letters or words as indicated.	tr	He the proof marked.
stet	Let it stand as it is. Disregard all marks above the dots.	stet	He marked the proof.
/=/	Insert hyphen where indicated.	/=/	He made the proofmark.
eq.#	Equalize spacing.	eq #	He marked the proof.
[or]	Move over to the point indicated.		
	[if to the left; if to the right]	[	He marked the proof.
⊔	Lower to the point indicated.	⊔	He marked the proof.
⊓	Raise to the point indicated.	⊓	He marked the proof.
ϡ	Insert comma where indicated.	ϡ	Yes he marked the proot.
℈	Insert apostrophe where indicated.	℈	He marked the boys proof.
ℰℰ	Enclose in quotation marks as indicated.	ℰℰ	He marked it proof.
H=	Replace with a capital the letter or letters indicated.	H=	He marked the proof.
sc	Use small capitals instead of the type now used.	sc	He marked the proof.
⊥	Push down space which is showing up.	⊥	He marked the proof.
⌒	Draw the word together.	⌒	He marked the proof.
⌢2	Insert inferior figure where indicated.	⌢2	Sulphuric Acid is H₂SO₄
2⌣	Insert superior figure where indicated.	2⌣	a² + b² = c²
Out, see copy	Used when words left out are to be set from copy and inserted as indicated.	Out, see copy	He proof.
æ	The diphthong is to be used.	æ	Caesar marked the proof.
ﬁ	The ligature of these two letters is to be used.	ﬁ	He filed the proof.
spell out	Spell out all words marked with a circle.	spell out	He marked the 2d proof.
¶	Start a new paragraph as indicated.	¶	reading. The reader marked
No ¶	Should not be a separate paragraph. Run in.	No ¶	marked. The proof was read by
? (circled)	Query to author. (Encircled in red.) This is the symbol used when a question is to be set. Note that a query to author is encircled in red.	was? (circled)	The proof read by
2.		?	Who marked the proof.
=	Out of alignment. Straighten.	=	He marked the proof.
/—/	1-em dash.	/—/	He marked the proof.
/2—/	2-em dash.	/2—/	He marked the proof.
/–/	En dash.	/–/	He marked the proof.
⊓	Indent 1 em.	⊓	He marked the proof.
⊓⊓	Indent 2 ems.	⊓⊓	He marked the proof.

Figure 18 Marks used in correcting proof. (Reproduced with permission from *Mathematics in Type,* The William Byrd Press, Inc., Richmond, Va., p. 39.)

script to proof than the most experienced technical editor or professional proofreader.

- Be on the lookout for typographical confusions such as the following:

Capital Z	Numeral 2
Capital S	Integral sign $\int$
Superscript[1]	Single quote ' or prime '
Lower case "ell"	Arabic numeral "one"
Upper case "oh"	Arabic numeral "zero"
Lower case i	Greek ι (l.c. iota)
Lower case n	Greek η (l.c. eta)
Lower case p	Greek ρ (l.c. rho)
Lower case u	Greek μ (l.c. mu) and Greek υ (l.c. upsilon)
Lower case v	Greek ν (l.c. nu)
Lower case w	Greek ω (l.c. omega)
Upper case X	Greek χ (l.c. chi) and multiplication sign $\times$
"Contained in" symbol $\in$	Greek ϵ (l.c. epsilon)
"Less than" symbol $<$	"Greater than" symbol $>$

The main source of such confusions is the poorly handwritten character in typescripts. A copy editor will identify each symbol when marking your manuscript for the compositor, but in reading the proofs you are more sensitive to the nuances of your special symbols than is the proofreader, who relies on comparison-checking against the manuscript.

- Examine the structure and headings in the tables. Complex tabular material presents difficulties to the compositor. Check the vertical and horizontal alignment of the numerals, and be certain that the headings are accurately positioned. Proofread the numerical values against your original data. (Details for the form of tables were given in Chapter 14.)
- Review the final form of the artwork. If charts and drawings were redone, check the accuracy of plotted curves and bar heights. See that all the callouts on charts and drawings are included. See that coordinate labels on graph charts are clear and specific. Inspect cropped photographs to ensure that they show all that you intended.
- If the printer's proofs that you receive are pages instead of galleys, see that the tables, figures, and captions are properly placed and numbered.

Accept the responsibility to proofread your own work carefully. Should an ugly mistake appear in your published report or paper, it will be a reflection on you, the author, and not on an unnamed proofreader at the printshop who happened to lack your insights into the manuscript.

Chapter 24

How to Present a Paper Orally

A Comparison: Written Versus Oral Papers

An oral presentation is far different from a paper to be published. If you plan to present your paper orally, remember the special limitations of your listeners. Whereas the reader of the printed page has the freedom to scan the headings and sections, to skip material at will, to proceed at any desired speed, and to re-read when necessary, listeners in your audience have no such freedoms. They depend on you, the speaker, to stake out a path to be traveled at a sensible pace. The expression "captive audience" is not a misnomer.

Because of these inherent differences, your written manuscript (which may be entirely appropriate for a conference proceedings or a journal) can fall flat when you attempt to read it to your audience word for word.

If in your oral delivery you choose to be "wedded" to the manuscript, two kinds of conflict arise that will be grounds for divorce. The first lies in the *pronounced* distinction between spoken and written English. The second results from the opposing purposes of oral presentation and formal writing. Your understanding of these distinctions will be reflected in the character of your talk.

The characteristics of the oral paper depend first on your voice. Your manner of speaking conveys certain impressions that do not come across in your writing. Voice inflections (changes in either pitch or loudness) give an added dimension to your paper, providing another way to emphasize the important sections and to subordinate the ideas in others. Your attitude is another parameter of speech that is quickly sensed by the audience. Enthusiasm about your subject is infectious, and lethargy is equally so. Use whatever words you may, you can never hide your own boredom from your listeners.

153

Even more important than speaking mannerisms are certain other aspects of the oral paper. These depend on your empathy for the audience, as contrasted with empathy for readers. Evidence of your feeling for the audience, for example, is the occasional gesture—a natural accompaniment of speech. When you are showing slides or foils and point to parts of a diagram on the screen, your explanations are direct and immediate. The text in a printed paper, on the other hand, is less direct when it refers to the figures. In that case the reader must look back and forth from figure to text to glean the details.

The amount of detail that is appropriate for slides may differ from what you provide in figures for your printed paper. The decision on how much detail to include depends on your sensitivity to what the audience can see and understand, particularly those seated in the rear. When the conference does not publish a proceedings for future reading, it is unfair to illustrate your talk with complicated slides, loaded with labels and legends. Not everyone in the audience will have a camera to photograph your slides for future study.

Still another aspect of the oral presentation is its flexibility. Unlike the written paper, the length and content can be adjusted according to feedback from the audience. When your listeners are shifting uneasily in their chairs or straining to see with drooping eyelids, it is time to speed up your presentation, to skip some of the details, and to put more life into your voice. When you see them sitting on the edge of their chairs, waiting to catch every word, you know it is time to slow down and perhaps give more details than you had planned. Be careful, however, to stay within the time slot scheduled for your presentation.

Usually your talk will be shorter than your published paper. This limitation is important: it emphasizes two more differences between oral and written versions. The first is that oral presentations have rather rigid time allocations, especially for papers given in panel sessions at a conference. If you overrun your time slot, you shorten the time available for a discussion period. Journal manuscripts are not so limited. The second difference is the attention span of listeners versus that of readers. In the oral version you must assume that your audience will tire of long-winded explanations and voluminous data. Your presentations should be short and to the point, illustrated where necessary on the projection screen. The written version, having additional supporting material, tends to be lengthier and to contain more data for future reference.

Publishing a report or journal paper has a special value. It becomes a matter of record; you become known for your work; your professional activity becomes recognized by your peers. Presenting a paper orally offers certain additional advantages. It is your opportunity to meet with engineers in your field, to trade notes on newly developed methods and on recent trends, and thus to broaden your horizons. And in the discussion period after your talk you also have the advantage of immediate peer reaction. Your presentation, then, is somewhat different from your published paper in the way it contributes to your professional development.

For these reasons, it is worthwhile to plan your oral paper separately, to design it for its special purpose, and to find out beforehand the makeup of your audience, as suggested by Barnow.[8]

Preparation of the Oral Version

There are three alternatives to reading your paper verbatim: glancing occasionally at a brief, topical outline; using a stack of note cards; or simply talking from slides, foils, or a flipchart. The last method—building your talk around visual aids—is probably the easiest. Because most engineering papers depend on charts, tables, or photographs for the central points, your audience anticipates that treatment and will be comfortable with an illustrated talk. Moreover, you will be using your visual aids as cue cards and can then *talk* to the audience without reading notes verbatim.

The first consideration for the structure of your talk is whether a written version will be available to your listeners either as a proceedings paper or a handout, or possibly as a future journal paper. If there will be no written version for reference, your oral presentation should be long enough to provide all the needed technical details.

The second consideration is the way you are to shape your talk for three overlapping purposes:

- to fit the technical program
- to meet your own objectives
- to match the interests of the audience

The program director or session moderator can let you know about the desired treatment and limitations of your subject and, most important, about the type of attendees and their average technical level. Into this reference frame you must fit the subject material that will fulfill your own purpose, whether it be to describe a system, solve a problem, develop a theory, or take sides in a controversy. Remember that your material, no matter how well presented, will be a disappointment to all (including yourself) unless it is tailored to fit the program and the audience.

When you are planning the various portions of your talk, deciding what to include is half the problem. Deciding what to omit is the other half. Assume, for example, that your presentation is limited to 20 minutes and you have a longer written version for the conference proceedings. The 15 typewritten pages, 250 words per page, contain a total of 3,750 words. If you were to read that version at an average speed of 125 words per minute, the time required would be 30 minutes. But in your 20-minute program slot you can use only two-thirds of the written content.

In this case you would need to cut at least one-third of the material. Remember also that listening is slower than reading; the listening audience cannot absorb as many details as readers. Therefore, it is best to delete portions that you feel will be less interesting and less suitable for the attention span of an audience. Omit the kind of detail that a listener, if sufficiently interested, can find in the proceedings. In addition, it is wise to allow extra time to define special terms in your talk. Listeners, unlike readers, cannot stop to look up definitions.

The most meaning in your presentation will probably be found in visual aids, which have a much stronger effect on the viewers than do the figures in a printed paper. A slide projected on the screen is a powerful magnet for attention because it is the only thing in view. One of your crucial decisions in using such a potent information display is your choice of illustrations. In any misguided effort to sustain interest with trivia or entertaining distractions on the viewing screen, you are defeating your main purpose.

Some of the more important figures in the written paper can be reproduced and made into slides or foils for your talk. However, any figure that seems too complex for projection on the screen should be redrawn and simplified. One criterion for simplification is the technical level of your audience: the lower the level, the less detail is appropriate.

If you build your talk around a series of slides, you will probably find it necessary to use connectives and liaison devices more liberally in your talk than in the written version. In the latter, some of the figures are probably not interrelated because they are separated by long sections of text. A smooth oral presentation should be heavily interspersed with expressions relating one slide to the next, such as

> In contrast to the sharp changes in the slopes of these curves, the next slide shows an increased stability . . .
>
> Another example of the functions in Figure 6 is shown in Figure 7, which illustrates the use of the system in . . .
>
> This photograph is an external view of the device, and the next slide is a line drawing indicating the structural details . . .

Such connective signposts help the audience stay on the road with you. If you do not use connectives, and especially if you switch abruptly from one topic to another, your listeners will begin to drop by the wayside. Some will be wondering where to go for lunch. Others will be looking over the list of speakers that follow on the program. Still others will quietly snooze.

To capture and maintain interest, then, your visual aids should be selected to illustrate important points, designed for the technical level of your particular audience, and arranged in a carefully connected sequence so that your presentation builds up interest to the end, when you offer conclusions or a summary.

After you have prepared the talk, rehearse it aloud. The best way, by far, is to use a tape recorder. Listen to the entire presentation. The defects in speech mannerisms will undoubtedly surprise you, because listening to yourself on the loudspeaker is far different from hearing your voice through the bones of the head as well as the ears. The feedback of your own word sounds provides you with an ideal opportunity to analyze your speaking habits. You also need an entirely different kind of feedback. Make a dry run of your talk for your colleagues and ask for their reactions to both speaking technique and technical content. These two trial runs are your insurance against a poorly prepared and badly timed presentation.

Delivering Your Talk

On the day of the talk, get to the conference room a few minutes early, equipped with your notes, slides, foils, charts, or other aids, carefully numbered and in proper order. In case there is no session moderator, room attendant, or audio-visual assistant, check the facilities yourself: projection equipment, spare projector bulb, microphone, position of lectern, and operation of room lights. Doing this before the audience is fully seated will save time and prevent confusion when you are ready to start.

As the starting time arrives, you may experience stage fright. Many people, even the most experienced speakers, have this problem during the first few minutes. The best way to face stage fright is to accept it as a natural reaction and to realize that the associated tension is not altogether undesirable. If you are too relaxed, you will be less alert and perhaps even less motivated to do an outstanding job as a speaker.

Nevertheless, you should find ways to minimize stage fright so that it does not get out of hand. This frustrating emotion is basically a fear of being inadequate when facing an audience. The obvious antidote is to bolster your self-confidence in several different ways. The first has already been discussed: prepare your talk thoroughly and fine-tune it for its purpose. In addition, you should make prior contact with members of the audience. Before your talk, strike up conversations individually with several audience participants. You will probably develop the feeling that they are on your side. You will sense that they want you to succeed. As a result you will be reinforcing your feeling that you have something important to say and that your presentation will be filling a need. When you ascend the podium, show your new confidence by glancing at the audience with a smile of anticipation. That attitude is always infectious, and the audience will tend to relax and return your smile. As a member of a "mutual admiration society," you should find your stage fright dissipating to a manageable level.

When you face an audience, you should be conservatively dressed. Loud colors or flashy jewelry are a distraction. Avoid, also, any posture or movement that distracts attention. Stand comfortably erect. Do not slouch over the lectern. If you do not know what to do with fidgety hands, let them hang loosely at your sides.

Other kinds of movement are not at all distracting but instead provide emphasis and variety to your presentation. Moving occasionally away from the lectern serves to stimulate audience interest when it is lagging. Natural gestures of the kind you ordinarily use in conversation will help drive home a point. (Dramatic and exaggerated gestures should be avoided.) Pointing to features on the screen or flipchart helps to focus attention and adds life to your talk.

When showing visual aids, never turn your back on the audience. It is best to maintain the all-important eye contact with your audience, which will in turn give your listeners the feeling that you are talking to them directly. Even in a large hall, an occasional direct glance at people in various parts of the room serves another purpose: the eye-to-eye rapport helps you to sense their reactions and take cues from their behavior.

Your command of the speaking situation on the podium is an intangible quality called *presence*. When you have developed an attitude of full commitment and self-confidence, and an interest in the audience as well as your notes, your presence becomes a strong asset to the speech.

Speech

Presenting your paper orally gives listeners the advantage of hearing your personal interpretations. This is especially true when you *talk* to the audience instead of mechanically reading your paper aloud. People are more comfortable with the easy flow of a semi-conversational style than with the tight constructions of formal writing. Both volume and pitch of your voice tend to rise and fall in the patterns of language. Your normal variations in speed also add shades of difference in meaning. These changes in volume, pitch, and speed are most useful but should not be overdone. Your speaking style must not be oratorical. The subtleties and nuances in ordinary speech will not be lost on your listeners.

Unless you have a great deal of experience in public speaking, you will need to have conscious control over parameters such as speed. If you are nervous when you begin to speak, start off slowly and deliberately before you build up to your normal speaking pace. The tendency is to talk too fast. The rapid speaker sometimes adopts a high "twangy" pitch, which generates tension in the audience, and is less easily understood than a slower voice in lower register, which has a more pleasing resonance. At rates higher than about 160 words per minute, your articulation begins to lose precision. To pronounce each word clearly, the tongue, lips, and teeth have to move freely and without tension. Unless you have better than average articulation skills, you had better stay with the medium speeds. The extremely slow speaker usually drones monotonously before an impatient and frustrated audience.

The least obvious but most effective way to vary your speed is to talk faster when you wish to arouse interest and a bit slower when you sense that a significant point needs to sink in and be understood. For most of us the typical range of speaking rates is 120 to 160 words per minute.

A loud voice is not necessarily the best attention getter. When you come to a part of your talk that deserves emphasis, try dropping your voice just a bit. It will be better to have your listeners leaning forward occasionally to catch every word than to subject them to a continuous blast. If you are using a microphone on a stand, speak in normal tones. When you abandon the microphone to discuss visual aids, remember to speak loudly enough to be heard in the rear of the hall.

You customarily adapt your speaking mannerisms to the subject and to the portion of the work you are discussing. The sections of the paper where you need to be the most sensitive to speaking style are those where audience interest is highest—the beginning and the end.

Openings and Closings

Speakers like an attentive and interested audience. When you start your talk, you are the center of attraction. Those seated before you not only expect you to bring your subject into clear focus but are also appraising your personality. When you seem to be approaching the end of your talk, they expect you either to summarize or, if you are to offer an analysis or recommendations, to give them the punch line. In openings and closings do not disappoint your audience.

A good way to begin is to come immediately to the point. Define your purpose, outline the scope of your paper, and show what you intend to develop.

Some speakers prefer at the very start to delay the prepared talk while they indulge in remarks that will place the audience at ease. An anecdote or a light-hearted aside can put listeners in a receptive frame of mind, but you can use witty or chatty openings successfully only if they are a natural expression of your personality. A straitlaced speaker who attempts a weak joke to get the audience in a good mood is off to a terrible start. A flamboyant speaker who opens with off-color humor can also derail the presentation.

Your best approach is to state your case and to establish good relations with your audience. If the talk will be in any way controversial, or if you happen to face a hostile audience, stake out some common ground. Be tactful. Remind them of what they already know; discuss the areas of agreement; point out the respects in which you are on their side; and then gradually bring into play your special point of view.

As you near the end of your presentation, you will probably make a remark like "In closing, I have shown . . . ," "I will now summarize . . . ," or "In conclusion, I recommend that . . ." At this point, listeners who had been fidgeting or dozing will suddenly look up. You have forced their attention.

A weak ending is deadly. Never give the impression that you are exhausted or apologetic, that you really have nothing more to say, and that you merely wish to thank the audience for its kind attention. Because the end of your talk makes a much stronger and more lasting impression than the beginning, do not hesitate to summarize your best points and to show why they are significant. Another way of strengthening your closing remarks is to acknowledge the limitations of your design or your results and to point out the aspects that are not yet understood and those that have to be studied or developed further. Such honest and candid comments tend to take the wind out of the sails of your critics.

The Discussion Period

If your talk is followed by a session of questions and comments from the audience, you have the opportunity to add information you might have omitted, to clarify points in question, and perhaps to defend your paper against criticism. You should be able to anticipate some of the questions beforehand and come prepared with answers. A good way to fortify yourself against stickler questions before the meeting is to have a dry run with colleagues, as suggested earlier in this chapter. At that time you can practice answering the probing questions of friendly hecklers.

The discussion period after your talk can be especially interesting as an indicator of how well you have reached your audience. The number of questions asked can be revealing. The lack of any queries at all should hint to you that, although your talk may have been valid, it probably did not interest your particular group of listeners or that you loaded it with too many facts. A large number of questions would suggest that you must have hit a quivering nerve. Although numerous questions do not necessarily mean that your paper has merit, excited comment means, at least, that your listeners have responded.

The character of the questions is also revealing. If the queries deal mostly with the main concepts and results in your paper, you may be assured that the audience is alert to the gist of your contribution. If, on the other hand, the questions concern inconsequential side issues, you may begin to wonder whether your central points and your real message ever reached your listeners.

Some of the audience may comment about their own experiences or offer observations that have no bearing on your paper. You have to tolerate "show-off" comments politely but should not let them deteriorate into pointless rambling.

When you receive a direct question about your talk, *repeat it for all to hear.* This, incidentally, gives you time to think about it. Respond with an honest and forthright answer. If you do not know the answer, admit it freely. When the question seems too complex or poorly framed, attempt to draw out the questioner. You may then find that a seemingly irrelevant question turns out to be a significant and interesting one that deserves an illuminating answer.

If the questioner raises a highly controversial point, do not allow yourself to be cornered into a defensive stand. Maintain your control of the situation by suggesting that you can discuss it privately without involving the audience in such a time-consuming matter.

When you respond to questions tactfully and with good judgment, the discussion period can be a fruitful exchange of information. Filling in the gaps in this way then becomes the finishing touch to your oral presentation.

To sum up, an oral presentation deserves a somewhat different treatment from a written paper,[28] because even the most mature technical talk is usually a grown-up version of "show and tell." It is shorter and has less detail than the written version because we learn slower with our ears than with our eyes. In addition, the speaker's personality and sensitivity to the audience are important elements of an oral presentation.

Chapter 25

Summary: How to Avoid Strategic Errors in a Manuscript

The Patterns of Error

Most of the chapters in this book include comments on writing and publishing strategies. Unlike a deficiency in grammar, style, or format, the strategic mistake is difficult or impossible for a technical editor to rectify. This summary chapter reviews the ways an author can avoid such errors in judgment.

In the overall patterns of error, engineering manuscripts suffer from three different kinds of mistakes and improprieties.

- *Writing style and mechanics:* Weak exposition and rhetoric; mistakes in grammar, spelling, punctuation, abbreviations, etc. (Refinements are made in successive drafts by the author or by a qualified technical editor.)
- *Technical inaccuracies:* Errors in the data, experimental technique, author's interpretation, etc. (These errors are usually best resolved by the author, sometimes with editorial help.)
- *Strategic errors:* Lack of a well-balanced plan; poor choice of co-authors; insensitivity to reader interest; bad timing; inadequate literature search; misplaced technical emphasis; lack of credibility; absence of colleagues' feedback; ignorance of referees' criteria; weak choice of a journal. (Errors of this kind are basic to an author's thinking. Usually questions of strategy must be resolved by the author.)

The following review and interpretation is intended to help you set up an overall plan and supporting strategies.

162

A Basic Strategy

My recommendation is to draw up a topical outline, like the one in Chapter 4, as a guide to your writing effort. Although as you write the manuscript you will probably be altering such a skeletal guide, its initial form will still have a powerful purpose: it will serve as a *strategic plan*.

Never think of the outline as being solely a device to set up the sequence and structure of your ideas. As pointed out in Chapter 4, the content of the outline should also strike a balance among three strong forces: your motivation for writing, the technical objectives of the work, and the interests of anticipated readers.

Of those three forces, to be harnessed in the manuscript as part of your grand strategy, potential reader interest is probably the strongest. Be certain that the topics in the outline relate both objectives and results to the interests of future readers in a meaningful way, as discussed at length in Chapter 2 and summarized in Figure 19.

As an important part of your basic strategy, adopt a technical level that will be understandable to your audience. When readers are at two separate levels (such as business executives and engineering specialists), use the strategic solution to that problem. In the introduction select topics for the level of the general reader, and in the other sections write for the level of the specialist. An alternative is to insert into your outline a section entitled *Summary for the Executive Reader* or *Executive Digest*. The topics in such a technical summary should be chosen so that it will not "talk down" to the executive. Rather, it should stress the business aspects and the overall technical significance.

A well-constructed outline, matching your objectives with the needs of the future audience, is the basis of a sound and effective plan. To complete the plan, you need to strengthen it further with several supporting strategies.

Supporting Strategies

Preliminaries

One decision that can have a strong influence on the manuscript is the choice of co-authors. Any of your colleagues who has made a major contribution to the work is, of course, a candidate for co-authorship, as mentioned in Chapter 3. The decision about having co-authors is, however, a strategic one. A co-author who writes well and is recognized as a contributor to the literature will be an asset. On the other hand, if you share the writing load with a poor writer, you may encounter no end of problems in completing the document. And if the manuscript is to be submitted to

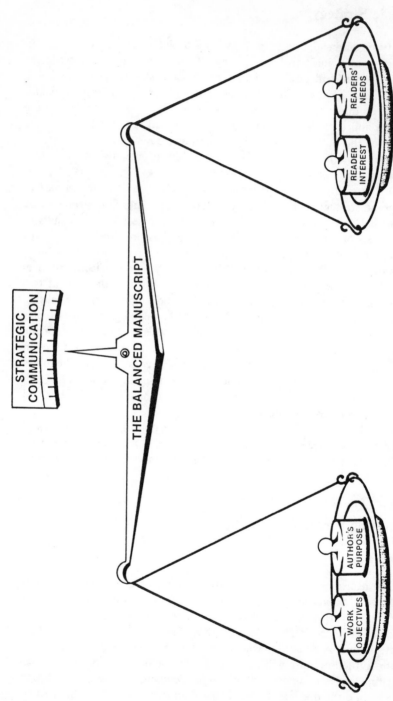

Figure 19 A strategy for effective communication. This well-balanced manuscript is constructed so that the work objectives and the author's purposes are evenly matched with reader needs and interests.

a journal, your co-author's poorly written sections might first require a good deal of rewriting. If circumstances require that you have a co-author with poor writing skills, you might tactfully suggest that he or she contribute to the manuscript only as planner and consultant.

As part of the planning, do not underestimate the value of the literature search. If you made one during the course of the engineering project, you are already equipped to show in your introduction how your paper differs from other published work. If you did not make a search, it is wise to do so before attempting to write a manuscript. Be aware of what has appeared in other engineering reports and journal papers. Your understanding of what others have accomplished in your specialty should be clearly evident in your manuscript. This is a protective strategy. It will shield you from accusations, later on, of wastefully duplicating the results of others and of being ignorant of engineering progress in your field. The techniques of the literature search are explained in Chapter 16. Referring to published papers is not a separate explanation, to be added to the manuscript; it is an inherent part of writing procedures.

Writing Techniques

The best writing strategy is to prepare a manuscript in a way that helps the progress of your engineering work. My suggestion of such a technique, described in Chapter 5, is useful if you are working on a development project and intend to prepare a report or paper for publication at the end of the project. Writing in increments as the work proceeds has two special benefits. First, you put your ideas on paper while they are still fresh—not as an aftermath of the project, when they will have become stale from repetitive thinking. Second, the insights you develop while writing will actually strengthen the course of your development work. In its final analysis, *writing helps the work*, and you should utilize this mechanism as part of your personal strategy.

Another strengthening aspect of writing techniques is the careful choice of topics to be emphasized and the methods stressing their importance. In the text your main contributions can be brought into sharp focus by the structural, visual, and rhetorical methods outlined in Chapter 10. These strategies include devoting more space to the central ideas and condensing the peripheral material; using headings and overall design as a means of highlighting information; and using the techniques of subordination, repetition, and comparison.

Making the right decisions on what to emphasize requires your critical judgment, not only of how much informative nourishment your readers will need but also of how much they will be able to digest.

In figures and tables similar principles apply. These are illustrated in Chapters 13 and 14. Visual aids, designed for their special purpose in the manuscript, should concentrate on topics that deserve emphasis. When their information content is poor, visual aids are not only weak but can also have negative effects. For example, confusing figures and tables, belabored with useless detail, are the result of a misguided effort to provide abundant information.

Anticipating the limitations of reader interest is one kind of strategy; spelling out the limitations of your results is another, as mentioned in Chapter 9. One way to add credibility to your manuscript is to point out candidly the pros and cons of your results. For example, the summary section of a paper on new gear designs might include statements of the limitations, e.g., "The new plastic gears will not operate at temperatures above 175°C; the reported stress analysis does not apply to loads higher than 2500 kg; and more work is to be done on development of glass-fiber-reinforced materials for higher impact resistance." The average engineering reader will have more confidence in concluding statements of this kind than in unqualified claims.

The opening sections of a technical manuscript are also especially susceptible to criticism. The first draft of the introduction is frequently weak and uninformative. A strategic approach will get your manuscript off to a good start and is especially important if your subject is controversial or if for any reason you anticipate a hostile audience. Play the role of a diplomat. Start by staking out areas of common agreement, as mentioned in Chaper 24. Define the problems to be discussed. Give a rationale for solving them. These initial statements will at least lay the groundwork for offering facts and arguments later to support your contention or proposal.

Another way to fortify your writing against criticism is to check your finished draft against a standard list of referees' criteria such as that given on p. 124. The questions listed were designed for use with journal papers, but they also apply to most kinds of engineering reports. After completing the draft, you will find such a checklist useful for finding omissions and other deficiencies in your manuscript. To get an even better perspective, test the waters. Ask a colleague to review the manuscript and give you critical comments based on the checklist.

If you receive negative criticism on your completed, edited manuscript—journal paper, technical report, conference paper, or thesis—try to evaluate the evaluator. Whether the critical comments are oral or written, try to analyze them along the lines suggested in Chapter 21. If objections are explicit, refute them directly or make appropriate changes in the manuscript. If the comments are general or vague, ask for specifics. In any case, review all the criticisms. Do they seem competent, unbiased, factual, and helpful? Or are they trivial, arbitrary, pointless, extremely general, or just plain wrong? Regardless, offer a polite rebuttal if an answer is required. The wise strategy is to defend your manuscript with the best facts you can muster, with additional interpretation, and without rancor.

If, on the other hand, you find that your critics were right and that your manuscript is an indefensible mistake, use the best strategy: withdraw it gracefully.

Choice of a Publication Vehicle

If your manuscript is intended to appear in print, you should include a definitive abstract and choose the best publication. The wrong choices can result in one or more of the following evils, all due to poor strategies on your part:

- Rejection by journal editors because of unsuitability for their readership
- Unnecessary publication delay of months or even years
- Burial in an obscure journal of small circulation
- Little or no recognition by your peers
- No citation of your work in other papers
- Poor retrievability in data banks

The way to avoid these difficulties, even before you write the manuscript, is to review the procedures given in Chapter 20. Look into the policies and reader interests of various journals. If you choose a trade magazine, call the editor as your first step, and send him an outline for review and comment.

The same strategy applies to other kinds of manuscripts. For in-house reports, you should first review a detailed topical outline with your supervisor. For a thesis, go over the entire plan with your thesis advisor. In either case it is your responsibility to ask about the technical areas that need to be emphasized and about the intended audience.

The segment of your audience that will grow rapidly in coming years is those who find papers on your subject by making an automated literature search. Be certain the abstract of your paper or report contains the key technical information described in Chapter 6. Otherwise your results will seldom be retrieved from data banks by those who need your information.

Oral Presentations

After you have written a paper, the decision also to present it orally gives you the opportunity to use an entirely different set of strategies. If your paper has new and interesting results, there are at least six strong reasons to present it at a technical conference or seminar.

The first is the opportunity for a "double play," i.e., giving it orally and offering the audience a written version, either in the form of a handout or a conference proceedings paper. The double exposure of your ideas is a distinct advantage. Moreover, there is a possibility for a third version of your paper because trade journal editors frequent engineering conferences, on the lookout to republish, with permission, short versions of attractive papers.

A second reason to offer an oral version is the opportunity to test your ideas on an audience. When giving your talk, you can quickly sense acceptance or rejection. The effects on the audience, either of excitement or boredom, will be unmistakable. As a speaker you can be like an artillery officer at a proving grounds. Fire away and observe the results.

Another strategic advantage of the technical talk is its flexibility. In observing audience reactions—either drowsiness or alertness—you can tell when it is time to skip certain passages or to expand others extemporaneously.

When presenting your paper, you have a certain strategic control over your audience that you do not have over readers. Any listener who is awake and attentive is following your presentation. Unlike a reader of hard copy, he has no chance to scan sections impatiently or to skip them altogether.

In the matter of emphasizing the important aspects of your work, you can bring some effective methods into play. The projection screen, for example, is a stronger hook for the attention than are figures and tables in a printed paper. In addition, the inflections of your voice provide shades of meaning and opportunities for emphasis that are unique to oral presentation.

These various reasons suggest that *reading* your paper verbatim from the lectern can be a drastic error. You have far more to gain by using the stratagems of the speaker who has eye contact with the audience.

The Bottom Line

In the ultimate sense, you have prepared a manuscript of quality only if it reaches its audience in a way that accomplishes your objectives. Strategic errors can mar that effectiveness even if the writing style and technical content meet your standards.

And so you must be a strategist as well as an author. To produce a fine manuscript describing your work is not enough. You need to understand the potential readership and relate your results to their needs. You must keep up with the technical developments in your field of engineering so that you can frame the manuscript in its proper perspective. Your plan for a sound document should also target its final form to reach readers within a reasonable time.

As an author you should understand the flow of engineering information and the ways to get your paper into the prevailing currents of the mainstream. Submitting your paper to the best journal with the largest circulation and the shortest lead time to publication then becomes one of your strategic efforts.

If you are part of an engineering organization that provides you with editorial assistance and information specialists, these professionals can help you with some of the strategic planning. But not all. The bottom line is the fact that you, the author, bear the responsibility.

Cited References

1. **American National Standards Institute.** 1977. *American National Standard for Bibliographic References.* ANSI Z39.29. American National Standards Institute, Inc., New York.
2. **American National Standards Institute.** 1979. *American National Standard for Writing Abstracts.* ANSI Z39.14-1979. American National Standards Institute, Inc., New York.
3. **Arms, V. M.** 1983. Engineers like to write—on a computer! *IEEE Transactions on Professional Communication* PC-26:175–177.
4. **Arms, Valarie M.** 1983. The computer and the process of composition. *Pipeline* 8:16–18.
5. **Arms, Valarie M.** 1984. The computer: an aid to collaborative writing. *The Technical Writing Teacher,* Spring 1984.
6. **Arnold, C. K.** 1962. The construction of statistical tables. *IRE Transactions on Engineering Writing and Speech* EWS-5(1):9–14.
7. **Ayer Press.** 1981. *Ayer Directory of Publications.* Ayer Press, Bala Cynwyd, PA.
8. **Barnow, R. K.** 1982. Audience analysis: Know who you're talking to—How, you say! *Proceedings, 29th International Technical Communication Conference,* p. C-1. Society for Technical Communication, Washington, DC.
9. **Barrass, Robert.** 1978. *Scientists Must Write,* p. 107. Chapman and Hall, London.
10. **Barry, J. G.** 1980. Computerized readability levels. *IEEE Transactions on Professional Communication* PC-23(2):88–90.
11. **Black, H. S.** 1934. Stabilized feed-back amplifiers. *Electrical Engineering* 53:114–120.
12. **Bowker Co.** 1982. *Ulrich's International Periodicals Directory.* R. R. Bowker Co., New York.
13. **Buehler, Mary Fran.** 1977. Report construction: Tables. *IEEE Transactions on Professional Communication* PC-20(1):29–32. *Also* Table design. *Proceedings of the 27th International Technical Communication Conference, 1980,* pp. G-69–G-73. Society for Technical Communication, Washington, DC.

14. **Cherry, L., and W. Vetterman.** 1982. Writing tools—the STYLE and DICTION programs. *Bell Laboratories Computing Science Technical Report No. 91.*

15. **Cole, Stephen, J. R. Cole, and G. A. Simon.** 1981. Chance and consensus in peer review. *Science* 214:881–886.

16. **Collier, Richard M.** 1983. The word processor and revision strategies. *College Composition and Communication* 34(2):149–155.

17. **Cremmins, E. T.** 1982. *The Art of Abstracting.* ISI Press, Philadelphia.

18. **Daiute, Collette.** 1983. The computer as stylus and audience. *College Composition and Communication* 34(2):134–136.

19. **Davidson, W. C.** 1977. Comment on "Protection of Authors: The Case for Anonymous Referees." *American Journal of Physics* 45(9):102.

20. **Day, R. A.** 1983. *How to Write and Publish a Scientific Paper,* 2nd ed., pp. 42–47. ISI Press, Philadelphia.

21. **DeBakey, Lois.** 1971. The intolerable wrestle with words and meaning. *Proceedings, 18th International Technical Communication Conference,* paper 9-2. Society for Technical Communication, Washington, DC.

22. **DeVolpi, A., G. S. Stanford, C. L. Fink, E. A. Rhodes, and M. R. Fenrick.** 1980. Text processing for the professional staff. *IEEE Transactions on Professional Communication* PC-23:164–167.

23. **Doherty, W. J.** 1979. The commercial significance of man/computer interaction. *Man/Computer Communication* 2:81–93 (Infotech State of the Art Report).

24. **Fluegelman, Andrew, and Jeremy Joan Hewes.** 1983. *Writing in the Computer Age.* Anchor Books, Garden City, NY.

25. **Gordon, Michael.** 1977. Evaluating the evaluators. *New Scientist* 73:342–343.

26. **Herzberg, A. M.** 1979. Note on the game of editormanship. *Physics Today* 9:11–12.

27. **Huth, E. J.** 1980. Does peer review of journal manuscripts improve the quality of scholarly journals? *Proceedings of the Society for Scholarly Publishing,* pp. 59–60.

28. **Institute of Electrical and Electronics Engineers.** 1980. Special Issue on Public Speaking for Engineers and Scientists (19 papers). *IEEE Transactions on Professional Communication* PC-23:5–52.

29. **Jones, F., and J. Paraszczak.** 1981. RD3D (computer simulation of resist development in three dimensions). *IEEE Transactions on Electron Devices* ED-28:1544–1552.

30. **Juhasz, S., E. Calvert, T. Jackson, D. A. Kronick, and J. Shipman.** 1975. Acceptance and rejection of manuscripts. *IEEE Transactions on Professional Communication* PC-18(3):177–185.

31. **Koefod, Paul E.** 1964. *The Writing Requirements for Graduate Degrees.* Prentice-Hall, Englewood Cliffs, NJ.

32. **Larsen, R. A.** 1980. A silicon and aluminum dynamic memory technology. *IBM Journal of Research and Development* 24:268–282.

33. **Lutz, Jean A.** 1984. A study of revising and editing at the terminal. *Transactions on Professional Communication,* pp. 73–77.

34. **MacDonald, N. H., L. T. Frase, P. S. Gingrich, and S. A. Keenan.** 1982. The writer's workbench: Computer aids for text analysis. *IEEE Transactions on Communications* COM-30:105–109.

35. **MacDonald, N. H.** 1982. The writer's workbench programs and their use in technical writing. *Proceedings, 29th International Technical Communication Conference,* p. W-70. Society for Technical Communication, Washington, DC.

36. **Masefield, John.** 1973. *Shakespeare and Spiritual Life.* Folcroft Library Editions, Folcroft, PA.

37. **Mathes, J. C., and D. W. Stevenson.** 1976. *Designing Technical Reports: Writing for Audiences in Organizations,* pp. 14–23. The Bobbs-Merrill Company, Inc., Indianapolis.

38. **McGirr, C. J.** 1978. Guidelines for abstracting. *Technical Communication,* second quarter, pp. 2–5.

39. **Michaelson, H. B.** 1950. Semantics and syntax in technical reports. *Chemical and Engineering News* 28:2416–2418.

40. **Michaelson, H. B.** 1959. Information gaps and traps in engineering papers. *IRE Transactions on Engineering Writing and Speech* EWS-2(3):89–92.

41. **Michaelson, H. B.** 1961. Creative aspects of engineering writing. *IRE Transactions on Engineering Writing and Speech* EWS-4(3):77–79.

42. **Michaelson, H. B.** 1961. Problems of style and semantics in technical writing. *Proceedings, Institute in Technical and Industrial Communications,* Fort Collins, CO.

43. **Michaelson, H. B.** 1968. Achieving a disciplined R and D literature. *Journal of Chemical Documentation* 8:198–201.

44. **Michaelson H. B.** 1974. The incremental method of writing engineering papers. *IEEE Transactions on Professional Communication* PC-17(1):21–22.

45. **Miller, L. A., et al.** 1981. Text critiquing with EPISTLE system, an aid to better syntax. *AFIPS Conference Proceedings* 50:649.

46. **Monroe, Judson, Carole Meredith, and Kathleen Fisher.** 1977. *The Science of Scientific Writing.* Kendall/Hunt Publishing Co., Dubuque, Iowa.

47. **Noyes, T., and W. E. Dickinson.** 1957. The random-access memory accounting machine. *IBM Journal of Research and Development* 1:72–75.

48. **O'Connor, M.** 1978. Standardization of bibliographic references. *British Medical Journal* 1(6104):31–32.

49. **Ogden, C. K., and I. A. Richards.** 1923. *The Meaning of Meaning.* Harcourt, Brace and World, Inc., New York.

50. **Pearsall, T. E.** 1969. *Audience Analysis for Technical Writing,* pp. ix–xxii. Glencoe Press, Macmillan, Beverly Hills, CA.

51. **Peters, D. P., and S. J. Ceci.** 1980. A manuscript masquerade. *The Sciences* 20:16–19.

52. **Qasim, S. H.** 1977. *SI Units in Engineering and Technology.* Pergamon Press, Elmsford, NY.

53. **Sherrington, A. M., and R. H. Orr.** 1965. The ethic of personal responsibility for scientific publications: Problems of authorship. *STWP Convention Proceedings,* Paper No. 23. Society for Technical Communication.

54. **Silver, Simon.** 1980. Ethical questions in the peer review system. *Proceedings of the Society for Scholarly Publishing,* pp. 67–79.

55. **Sternberg, David.** 1981. *How to Complete and Survive a Doctoral Dissertation.* St. Martin's Press, New York.

56. **Stibravy, John A., and Charles E. Beck.** 1984. The question of quality: New user attitudes toward word processors. *Proceedings, 31st International Technical Communication Conference,* pp. RET-31–RET-33. Society for Technical Communication, Washington, DC.

57. **Strunk, William, Jr., and E. B. White.** 1979. *The Elements of Style,* 3rd ed., p. 23. Macmillan Publishing Co., New York.

58. **Sugden, Virginia M.** 1973. *The Graduate Thesis.* Pitman Publishing Co., New York.

59. **Tichy, H. J.** 1966. *Effective Writing.* John Wiley & Sons, Inc., New York.

60. **Tracey, J. R.** 1966. The effect of thematic quantization on expository coherence. *IEEE International Convention Record,* Paper No. 9.4, Part II, p. 17.

61. **Tukey, J. W.** 1955. Systemization of journal practices (references). *Science* 22:246.

62. **Turabian, K. L.** 1973. *A Manual for Writers of Term Papers, Theses, and Dissertations,* 4th ed. University of Chicago Press, Chicago.

63. **U.S. Department of Commerce and R. A. Hopkins.** 1977. *NBS Metric Guide and Style Manual.* American Metric, Tarzana, CA.

64. **Weisman, H. M.** 1980. *Basic Technical Writing,* 4th ed. Charles E. Merrill Books, Inc., Columbus, OH.

65. **Williams, J.** 1977. Quality in the review process. Third IEEE Conference on Scientific Journals, 2 May 1977, Reston, VA. *IEEE Transactions on Professional Communication* PC-20(2):130–131.

66. **Zinsser, William.** 1983. *Writing with a Word Processor.* Harper and Row, New York.

Additional References

American National Standards Institute. 1979. *American National Standard for the Preparation of Scientific Papers for Written or Oral Presentation.* ANSI Z39.16-1979. American National Standards Institute, Inc., New York.

Andrews, D. C., and M. D. Bickle. 1978. *Technical Writing: Principles and Forms.* Macmillan Publishing Co. New York.

Blicq, R. S. 1981. *Technically Write: Communication in a Technological Era,* 2nd ed. Prentice-Hall, Inc., Englewood Cliffs, NJ.

Brunner, Ingrid, J. C. Mathes, and Dwight W. Stevenson. 1980. *The Technician as Writer: Preparing Technical Reports.* Bobbs-Merrill, Indianapolis.

Fear, D. E. 1981. *Technical Communication,* 2nd ed. Scott, Foresman and Co., Glenview, IL.

Gibaldi, Joseph, and Walter S. Achtert. 1977. *Handbook for Writers of Research Papers, Theses and Dissertations.* Modern Language Association of America, New York.

Government Printing Office. 1982. *Style Manual.* U.S. Government Printing Office, Washington, D.C.

Harkins, Craig, and D. L. Plung. 1981. *A Guide for Writing Better Technical Papers.* The Institute of Electrical and Electronics Engineers Press, New York.

Houp, K. W., and T. E. Pearsall. 1984. *Reporting Technical Information,* 5th ed. Glencoe Publishing Co., Encino, CA.

Kirkman, John. 1980. *Good Style for Engineering and Scientific Writing.* Pitman Publishing Co., Ltd., London.

Mali, Paul, and Richard W. Sykes. 1984. *Writing and Word Processing for Engineers and Scientists.* McGraw-Hill Book Co., New York.

Mills, G. H., and J. A. Walter. 1978. *Technical Writing,* 4th ed. Holt, Rinehart and Winston, New York.

Mitchell, J. H. 1968. *Writing for Professional and Technical Journals.* John Wiley & Sons, Inc., New York.

Pearsall, T. E., and D. H. Cunningham. 1982. *How to Write for the World of Work,* 2nd ed. Holt, Rinehart and Winston, New York.

175

Rathbone, R. R. 1966. *Communicating Technical Information.* Addison-Wesley Publishing Co., Inc., Reading, MA.

Sherman, T. A., and S. Johnson. 1975. *Modern Technical Writing,* 3rd ed. Prentice-Hall, Inc., Englewood Cliffs, NJ.

Shulman, J. J. 1980. *How to Get Published in Business/Professional Journals.* American Management Association, Inc., New York.

Sides, Charles H. 1984. *How to Write Papers and Reports About Computer Technology.* ISI Press, Philadelphia.

Souther, J. W., and M. L. White. 1977. *Technical Report Writing,* 3rd ed. Holt, Rinehart, and Winston, New York.

Teitelbaum, Harry. 1975. *How to Write Theses.* Simon and Schuster, New York.

Ullman, J. N., and J. R. Gould. 1972. *Technical Reporting,* 3rd ed. Holt, Rinehart and Winston, New York.

University of Chicago. 1969. *Manual of Style.* University of Chicago Press, Chicago.

Warren, Thomas. 1978. *Technical Communication: an Outline.* Littlefield, Adams and Co., Totowa, NJ.

Weil, Ben H. 1975. *Technical Editing.* Greenwood Press, Westport, CT.

Weiss, E. H. 1982. *The Writing System for Engineers and Scientists.* Prentice-Hall, Inc., Englewood Cliffs, NJ.

Woelfle, R. M. 1975. *A Guide to Better Technical Presentations.* The Institute of Electrical and Electronics Engineers Press, New York.

Index